Spacetime, Perception, and Reality

Reconsidered

Also, by this Author

Consciousness and Deity Reconsidered

History (and other things) Reconsidered

Space, Time, and Deity Redux

Spacetime, Perception, and Reality Reconsidered

Reconsidered

Proposing New Notions About Spacetime,
Perception, and Reality

John H. Silver

Beagle Publishing 3-7A

Contents

Introduction

This book is primarily the result of two incidents that occurred in the early 2000s. The first was a severe personal tragedy that compelled me to seek answers concerning life and purpose, and the second was the fortuitous encounter with the English philosopher Samuel Alexander's book *Space, Time, and Deity*[1] that is a two-volume compilation of his Gifford Lectures of 1916 - 1918.[2] This encounter was fortuitous because since the first incident, as mentioned, I had been searching for some answers to questions about the universe, life, and all that – but like so many others I found that the traditional answers did not fully satisfy.

*   *   *

This book is divided into two parts *Part 1: Time and Space* is a consideration and reconsideration of time and space. *Part 2: Perception and Reality,* similarly, is a consideration of what I call common perception and reality that is followed by a reconsidered or reassessment of both perception and reality. Both Part1 and Part 2 have new and revised material, and both parts provide background and supplemental material that should prove useful.

---

[1] Samuel Alexander (1920), *Space, Time, and Deity*. The Gifford Lectures at Glasgow. https://archive.org/details/spacetimeanddeit00alexuoft>.

[2] These lectures have been delivered annually since 1888 and here is what the Gifford Lectures website says: "The prestigious Gifford Lectureships were established by Adam Lord Gifford (1820–1887), a senator of the College of Justice in Scotland. The purpose of Lord Gifford's bequest to the universities of Edinburgh, Glasgow, St. Andrews and Aberdeen was to sponsor lectures to 'promote and diffuse the study of Natural Theology in the widest sense of the term—in other words, the knowledge of God'." Since the first lecture in 1888, Gifford lecturers have all been pre-eminent thinkers in their respective fields [https://www.giffordlectures.org/].

As the word spacetime implies, time and space are interconnected in such a manner that they are in some way inseparable and are part of our perceived reality.

In some ways, conceptualizing time has always been more challenging and problematic than space. In the case of space, we have sensual perception to assist us. We do not see space, but we do see that there is a separation between objects, both the separation of small objects nearby and the separation of heavenly bodies. This separation, the gaps between visible objects, is what we commonly call space.

Part 2 will show that perception and reality - more than merely related - are inextricably connected such that one is dependent on the other. Perception contains both objective and subjective elements to either greater or lesser degrees and neither element can be completely eliminated such that perception is always a mixture of the objective and subjective. To be sure, there are unalterable realities, but each person modifies and 'overlays' hard reality with a softer reality of their own making.

On the other hand, *time*, a thing not directly perceived in any physical way by our sense perception, is an abstract mental construction that we use to explain the separation of events; in a manner, somewhat analogous to the way we use space to explain the separation of objects. The obvious difference is that the separation of objects is directly observed, while our sensation of time is not directly observed and is variable to the extreme.

To assist the reader in understanding something about the nature of these speculations, brief discussions about past and present scientific theories are presented in as non-technical manner as possible, while still providing understanding of the science that has been developed during last century or so.

There have been fundamental changes in our understanding of the universe in the one hundred plus years since Alexander delivered these lectures, and consequently, his lectures are more like a starting point. His ideas were formed after the first great 20th century revolution in science - the theories of relativity were constructed - but before the second great revolution in science – quantum theory – was constructed and it is important remember that when Alexander's gave these lectures the universe was thought to be very small compared to our current knowledge. However, although somewhat outdated, Alexander's lectures were important to me as they jump-started my ideas about space and time that eventually led to this book.

To be sure, cosmology and other related observational disciplines have many questionable notions, however, enough is firmly established that we can state with confidence that new knowledge about the vastness of the observable universe suggests, or even requires, that we reconsider our notions about space and time.

Since the beginning of the Industrial Revolutions of the 18th and 19th centuries, there has been a trend away from traditional religion and toward progressive secularism where the new idols are the machines and as of today especially the computer in all of its many incarnations. This trend continued throughout the 20th century and accelerated in the early 21st century.

This acceleration has been aided by the technology that produced personal communication devices and made ubiquitous and addictive social media possible. One result of all this has been the widening of the gap between believers in nonphysical reality and those who do not such that there is, in some incidences, open hostility between those with opposing views. How this came about is beyond the scope of this modest work and, accordingly, the goal herein is to offer hints about a possible path away from overdependence on technology.

Until about a hundred years ago, the universe was considered to be an expanse of only a few thousand light years and was composed only of what we now call the Milky Way galaxy. Today, the known size of the observable universe has increased about a million-fold and is now measured in billions of light years. The evidence for this increase is strong but what is not very strong are some of the many notions that grew out of this new knowledge.

Current knowledge of the vastness of the universe and the unexpected strangeness of the micro-world as revealed by quantum theory require that we rethink some of our most strongly held beliefs. We need to remember that many of our beliefs originated thousands of years ago, in the setting of a small universe, and where the earth was the center of that universe.

As this work progresses, it becomes increasingly conceptual and speculative. In this way, I have attempted to open new notions about our world, which provides enrichment and a greater appreciation for the complexity and wonder of our world.

It is safe to say that some things that now seem complex will be simplified and some things that seem simple will become more complex. Portions of this book may be common knowledge to some, new and novel to some, and, possibly, completely absurd to some [hopefully not too many will have the latter experience]. Of course, the same reader could have all three experiences.

As a last word, I would like to thank my wife for her patience during the many hours I spent sequestered while working on this project.

Novato, California
October 2025
jhs8487@gmail.com

"In every investigation, in every extension of knowledge, we're involved in action. And in every action, we're involved in choice. And in every choice, we're involved in a kind of loss, the loss of what we didn't do. We find this in the simplest situation. ... Meaning is always obtained at the cost of leaving things out... In practical terms this means, of course, that our knowledge is always finite and never all encompassing. ... This makes ours an open world, a world without end."[3]

"The Red Queen shook her head. "You may call it 'nonsense' if you like," she said, "but I've heard nonsense, compared with which that would be as sensible as a dictionary!"[4]

"About the nature of the external world we have at the outset nothing but hypotheses. Before we test them in any very exact way, we may with safety try to understand them. Perhaps what seemed the wildest of them all may turn out to be the very best."[5]

"Talent hits the target that others miss, genius hit targets others cannot see."[6]

"The point of philosophy is to start with something so simple as not to seem worth stating, and to end with something so paradoxical that no one will believe it."[7]

[3] J. R. Oppenheimer, "The Oppenheimer Years," Los Alamos Science Winter/Spring (1983) http://la-science.lanl.gov/lascience07.shtml, 18.

[4] Lewis Carroll (1871), *Through the Looking-Glass, and What Alice Found There* (New York: Barnes and Noble, 2012), 133.

[5] Josiah Royce (1882), *Mind and Reality,* https://archive.org/details/mindreality00roycrich, 6.

[6] Ancient Chinese Proverb.

[7] Russell, Bertrand (1918), *The Philosophy of Logical Atomism*: London and New York: Routledge Classics, 2010.

Part 1: Time and Space

Considering Time and Space

Ask someone 'What time is it?' and the response will be without hesitation. However, ask 'What is time?' and you will get a very different response. That is, you are more likely to get a blank stare than you are an answer. Just to be sure, I tried this question on several unsuspecting victims, and received reactions ranging from suspecting a trick, to a perplexed look, and even a somewhat hostile reaction. After a quick explanation that the question is not a trick or a challenge, the unsuspecting victims usually relax.

Anyway, here are two examples of the more creative answers to the 'What is time?' question:

"I think I would say that time is the dimension required for there to be change and to measure change. And I think it is ultimately incremental, not continuous."

"Perceived as a continuum that passes in our lives, time is a static dimension that life passes through."

Since the owners of these quotes will recognize them – and just to be on the safe side - I won't make any negative comments or any comments at all. Well, maybe one or two comments. Although very different answers, both have a similarity in that each managed to avoid the circularity of using the word *time* to define time. Both responders also used the word *dimension*, but they disagreed on whether time is a continuum (smoothly connected) or discrete (non-continuous, incremental). Later, we will decide which view is the correct one.

To be sure, this question about time is not a new question such that we may reasonably assume that this question has been around so long that its origins are lost in the distant past. In fact, over the last two thousand years or so, we find that this question has been asked many times by individuals both famous and not so famous. For example, a classical and well-known example dating from the 5th century is a quote by St. Augustine (Christian-Roman Theologian, 354 – 430), which goes something like this; '*What then is time? If no one asks me, I know: if I wish to explain it to one that asketh, I know not.*'[8] A lot of us can relate to that, not only regarding time, but also to other common things of everyday experience.

The ancient Greeks, ever practical, are said to have seen time as a man riding backwards on a horse; he knows the past, but not the future.

Of course, not just time has been an ongoing topic of discussion throughout recorded history. Time has a companion called space that, over the centuries, has also been the topic of much discussion and debate. Anyway, of the two, time is more common in ordinary daily human activities. We use the word time in many ways, but seldom, if ever; do we think about what the word really means? Standard dictionary definitions are only somewhat helpful since they usually define the word time by means of everyday usage and often in a circular fashion. Dictionaries tend to use phrases such as *on time, good (or bad) timing,* and other phrases that are part of our normal daily conversations.[9]

Common Time

"So long as we do not go outside the domain of consciousness, the notion of time is relatively clear. Not only do we distinguish without difficulty present sensation from the remembrance of past sensations or the anticipation of future sensations, but we know perfectly well what we mean when we say that of two conscious phenomena which we remember, one was anterior to the other; or that, of two foreseen conscious phenomena, one will be anterior to

---

[8] St. Augustine of Hippo (354-430), *Confessions of St Augustine Book XI* (New York: Modern Library, 1999), 123.

[9] For example, in a particular dictionary I found 64 usage definitions for time, and 24 for space.

the other." [10]

"Time is what keeps everything from happening at the once!" Anonymous

"Time flies when you are having fun!" Anonymous

As stated in the quote by Henri Poincaré (French mathematician and philosopher, 1854–1912), if we stay within *the domain of consciousness* there is nothing uncomfortable about the notion of time. This is the common time of everyday experience, as influenced by past memories, current events, and the anticipated future.

Common time (aka naive time) is the time of our perceptions, and it is a learned intuition of psychology and not the *a priori* intuition of philosophy. That is, common time is a learned concept and not a natural intuition.

Put another way, common time is the time we experience in our everyday lives where, although we depend on clocks keeping a steady pace, our perception of this pace can vary greatly. For example, it is trivially obvious that we are all familiar with the difference between an hour spent under stress and an hour spent pleasantly entertained.

Natural Time

"Human perception is directed to the world; animal perception is directed to an environment."[11]

"Non-rational creatures do not look before or after but live in the animal eternity of a perpetual present; instinct is their animal grace and constant inspiration; and they are never tempted to live otherwise than in accord with their own animal dharma, or

---

[10] Henri Poincaré, (1898), "*The Measure of Time.*" en.wikisource.org/w/index.php?title=The_Measure_of_Time&oldid=2625853

[11] Merleau-Ponty 1964, 40.

immanent law."[12]

Imagine a world without clocks or calendars. No hours, days of the week, months, or any of the other many constructions and artifacts we use for keeping track of what we commonly call time. Imagine a world where life on earth was much simpler than today – just the earth with its daily passage of the sun, the cycles of the moon, the annular seasons, and the nightly spectacle of innumerable stars that appear to rotate about a fixed point. This is the world run by natural cycles and is the world of natural time. This *time* is unfettered by human influences such that it is the world of natural creatures today and that of humans in the dimness of prehistory, before becoming obsessively preoccupied with time.

Consider the following:

In the dead of winter in Antarctica hundreds of male Emperor Penguins will spend about 64 days huddled together while enduring fierce gales and frigid temperatures. Each male penguin does this for the sole purpose of incubating a single egg. During this period of sunless days, these male penguins have no food, and they will lose almost half their body weight!

How are the penguins able to do this? Put simply, they can do this because their sense of time is very different from our sense of time. Driven by natural cycles, which give them independence from the detailed schedules followed by most humans, they have a sense of natural time that is free from the anxieties of an anticipated future and regrets of the past. Put another way, as the second of the above quotes states, animals live in the present, while, on the other hand, humans live in the remembered *past* and anticipated *future* which are separated by a fleeting duration that we sometimes call the *present*.

---

[12] Aldus Huxley, *The Perennial Philosophy* (New York-London: Harper Modern Classics, 2009), 140.

For humans, such a task would be impossible. Even if we could endure the frigid weather and the lack of nourishment, our sense of time would render us incapable. We live by the clock, in a state of almost constant anticipation, and would soon find the sense of slowly passing time unbearable. On the other hand, instinct and the *Force of Necessity* are the drivers of the natural world, not clocks. *In the natural world, there is great patience.*

The natural cycles of day and night along with the lunar cycles and annular seasons have considerable influence on the natural world. However, this is not a human schedule driven by a clock, but is the union of perceived natural cycles with the creatures' physiological and psychological needs. In other words, natural time is *event-driven*, while our world is *schedule-driven*. We humans divide the day into hours, minutes, and seconds, then impose a schedule on ourselves that has little or no regard for natural cycles.

All life forms have an innate capability that *is a guide for appropriate action*. That is, all life forms possess, at a minimum, capabilities necessary for survival. Survival dictates that all life forms must interact and alter their immediate environment in pursuit of sustenance, reproduction, and ultimate survival. This is driven by the Force of Necessity such that pursuit of survival encompasses all living organisms from the great creatures of the ocean depths to the smallest microbes. To say that all life goes on within an environment may seem obvious but within our modern environment, this is often ignored because the natural environment and the true sources of sustenance are filtered out by the system of supply that provides us with products that bear little, if any, resemblance to their sources.

The interrelationship between the environment and humans is, obviously, quite different from that of the other creatures. Although all living creatures rely on a compatible environment for sustenance, it is only the human creature that imposes their will and makes unnatural demands on the environment. All other creatures work in and around their environment, while humans challenge and attempt to modify the environment with an attitude that seems to imply that *they* are the masters and not the natural world. This attitude is only adjusted or temporarily suspended by natural disasters and other natural occurrences over which we have no control.

Common Space

"Space is the boundless, three-dimensional extent in which objects and events occur and have relative position and direction. ... The concept of space is considered to be of fundamental importance to an understanding of the physical universe. However, disagreement continues between philosophers over whether it is itself an entity, a relationship between entities, or part of a conceptual framework."[13]

"Space: The physical universe beyond the earth's atmosphere. The near vacuum extending between the planets and stars, containing small amounts of gas and dust." [14]

"With Kant, space is given as a ready-made form of our perceptive faculty, a veritable 'deus ex machina'[15], of which we see neither

---

[13] Wikipedia definition.

[14] Online Oxford English Dictionary definition.

[15] *Deus ex machina* is a theatrical device frequently used by ancient Greek dramatists. It is a technique that provides for solutions to seemingly insolvable situations by the introduction of some arbitrary intervention by a god-like character that appears out of nowhere. It translates as *God from a machine*, which refers to the crane or other mechanical device used by the Greeks to make the *god-like* character suddenly appear. Today, this technique is often found in science fiction and fantasy writing. For example, J. K. Rowling's *Harry Potter* novels frequently use this dramatic device.

how it arises, nor why it is what it is rather than anything else...”[16]

Most people who have every thought about it probably think of space in a manner similar to that specified in the first two of the above quotes. Many, if not most, consider space as being a three-dimensional extension in which things exist and events occur. That is, space is seen as something like a container or box. I will call this *common space* and as with common time, we feel that we have an intuitive understanding of common space.

However, common space is different from common (linear) time; such that our notion of common space is a cooperative composite of individual sensory spaces. These individual spaces include visual, tactile, and motile spaces, each with a different sensory interpretation of space but all three are integrated into our overall notions of common space.

Visual Space is the space of the images produced by our sense of sight and is a two-dimensional projection on the retinae of our eyes. These images, having only height and width, lack a third dimension. The additional dimension of depth is a mental construction using the convergence of sensations in our two eyes.[17] That is, our perception of the third dimension (depth perception) is a mentally constructed interpretation.

Tactile Space is the space of our sense of touch. It assists in producing our notions about the reality of extended (substantive) matter[18] and contributes to our impression of the world being three-dimensional. Our tactile sense provides the

---

[16] Henri Bergson, *Creative Evolution* (London: MacMillan and Co, 1911), 216.

[17] Herein, I define sensations as the initial detection of external stimulus by a sense organ.

[18] Matter, as defined herein, includes not only extended persistent objects, but also includes energy and matter in all their known forms as well as any hidden structure and composition not yet discovered. Perhaps it should just be called the *stuff of the universe* or simply *world-stuff*.

sense data for building tactile space through a combination of the texture of objects and the feel of extension of these objects. For example, writing this sentence involves contact with the pen I am holding such that I am aware of the pen's texture and extension. Tactile space also plays an important role in helping us to establish the reality of objects. To touch an object is to believe it is real.

Motile (or Motor) Space is the third space that we learn as infants. We slowly learn to control the movements of our body parts, starting with jerky and uncertain movements we manage to progress, through a process of trial and error, to a level of proficiency that enables controlled and purposeful movement. This space not only helps us to move about and manipulate objects, but also, in conjunction with visual space, helps us build a sense of direction and awareness of our immediate spatial environment.

Common space is an integration of these sensory spaces, which enables us to function in what is primarily a two-dimensional environment. Sure, we sometimes go up or down, but for the most part, we move in two dimensions with the third dimension being generally limited to providing perspectives of size and distance, which helps us avoid walking into walls or other things.

Not only are the senses our window into the external world, but they work together to confirm our belief in veridical reality. For example, when we observe a simple cylinder, such as a can of beans, from one perspective we observe a rectangle, while from another we see a circle. Through rotation we observe the transition from circle to rectangle, and back again. The confirmation of the cylinder comes from our integration of the tactile sense, such as feeling the roundness and through visual sensations, assisted by the motile sense, which allows us to manipulate objects to gain a more complete image.

As we progress from infancy to maturity, our perceptual conceptions of space evolve and expand into our notion of common space. In other words, we build our notions of common space through the trial and error of perceptual experience.

Background Stuff

These next few topics are here to provide some information that should prove useful as background preparation for what follows. Some titles of these topics may seem technical, but they are presented in non-technical fashion to facilitate conceptual understanding by the non-specialist.

Classical Science

Modern science, or science as we think of it today, began developing around 1600 through the efforts of a few enlightened individuals such as Galileo Galilei (Italian, 1564-1642) and Johannes Kepler (German, 1571-1630). Although both remained partially under the influence of the ancient practices of alchemy and astrology, they were capable of going beyond these ancient occupations and introduced analytical elements into their investigations.

Their efforts were the beginnings of modern science. Kepler, for example, systematically analyzed a vast amount of astronomical data and developed kinematic theories about planetary motion that are still valid.[19] Complementary to the work of Kepler, Galileo pioneered modern scientific methods through the innovation of combining experiments with mathematics. Following the examples of Galileo and Kepler, others continued and enhanced the work they started.[20]

It was this period that created the classical image of scientists as being purely objective in their investigations and that

---

[19] Kepler used measurements collected by Tycho Brahe (Danish astronomer, 1546 – 1601), who for about 30 years recorded a vast number of celestial observations.

[20] The reader may have noticed that it is traditional to refer to scientists and other famous people by their family (last) name – such as Kepler or Newton - with one exception – Galileo is always referred to by his given (first) name. How this came about is another mystery to be solved.

the theories of science were equally objective. This notion, of course, was never completely true. Even though scientists were not considered part of experimental processes, there was always a personal aspect of investigations based on the scientist's inclinations, objectives, and agendas (possibly hidden). Therefore, even though mental aspects were not considered part of science and, while the experiments themselves may have been objective, the conjectures and hypotheses based on these experiments necessarily contain subjective elements influenced by the scientist's personality.[21]

In any case, a major concern of classical science was force acting on matter, also known as dynamics. Dynamics was introduced into modern science by Isaac Newton (English Physicist, 1642-1727) who, utilizing the kinematics of Kepler, brought gravity into the study of planetary motion.

Although Newton wrote extensively about gravity, he never attempted to determine its source, or at least, did not publish any such attempt and it was not until the 19th century that gravity began to be recognized as an intrinsic property of matter, and it was conjectured that gravity is a *field* propagated by some unknown mechanism. Put simply, in the scientific sense, a field is something that propagates between objects and has variable influence on these objects [See *Side Comment on Fields* below].

After gravity, the second fundamental field discovered was the electromagnetic field (EM Field).[22] This was the work of James Clerk Maxwell (English, 1831-1879), who built on the

---

[21] Herein, the *objective* is considered a reference to something independent of a perceiving mind, while the *subjective* is somewhat the opposite and refers to some notion held in the thinking mind. Put another way, objective is what we believe as having independent reality and subjective is a personalized interpretation or belief.

[22] The EM field could be considered the third field since the magnetic field has been known since antiquity – although it was not until the 19th century that it was recognized that magnetism was associated with *electrodynamics*.

earlier work of Johann Carl Friedrich Gauss (German, 1777-1855), Michael Faraday (English, 1791-1867), and others. Put another way, Newton's dynamics supervened on Kepler's kinematics, and Maxwell's electrodynamics supervened Faraday's electrostatics.[23]

Then in 1896, just as everyone was getting comfortable with notions about electricity and the electromagnetic field, along came the discovery of radioactivity. This was especially surprising and disturbing, because at that time the concept of the atom, although still being debated, was thought to be indivisible. The discovery of radioactivity – more or less - settled the debate about atoms, but more importantly, it radically changed the notion of an atom from something monolithic and indivisible into something with parts, and sometimes one or more of these parts are *kicked out* of the atom!

{**Side Comment on Fields**: A semi-technical definition of a field goes something like this; *"a field is a structure of continuously varying functions of position and time that are usually described mathematically using differential equations."*[24] Or, as Rischard Feynman (American Physicist, 1918–1988) put it; *"Physicists use the word 'field' to describe a quantity that depends on position in space and time."*[25] The most commonly encountered fields are the electromagnetic field and the gravitational field.

The electromagnetic field is conjectured to be propagated by microscopic bundles of energy called photons. On the other hand, it has been proposed that the gravitational field is propagated by a particle called *graviton*, which, if it exists, has eluded all attempts at discovery.}

---

[23] Forces that we know about have some theoretical propagator associated with them, even if this propagator has yet to be *discovered*.

[24] Wolfgang Pauli, *The Theory of Relativity* [Supplemental Note 23] (New York: Dover Publications Inc., 1981), 142.

[25] Richard Feynman, *QED: The Strange Story of Light and Magnetism.* (Princeton University Press, Princeton, NJ, 1985), 123.

{**Side Comment on the Electromagnetic Field**: It was the 19th century that saw great progress in examining and understanding what eventually came to be called the electromagnetic field. As the term implies, this field is a relation between a magnetic field and an electric field such that each can induce the other through relative motion.

As a magnet moves relative to a closed loop of metal somehow the magnetic field induces an electric current consisting of electrons in the metal loop moving in a certain direction and in quantities proportional to the strength of the magnetic field, while, on the other hand, an electric current (electron movement through a wire) such as that produced by connecting a wire to a battery induces a magnet field.

The concomitance between electric charges and magnetic fields are understood primarily only by the observed behavior and not through great insight into the actual causes of this relationship. There can be no rational doubt that the rotating magnetic field in a generator reaches out through space to induce electrical current but even after being studied and in practical use for generations we still do not really know why this actually happens. That is, we do not know exactly what causes a magnetic field except that it seems to have something to do with the inherent magnetic tendencies within certain metallic atoms.

In keeping with the propensity to attribute fields to specific field generating particles, propagators of the magnetic field have been proposed along the same lines as the postulated photons of electromagnetic radiation (aka light); however, the conjectured propagator of magnetism has not been found. It is just another one of the great mysteries of nature that for all our assumed intellectual capacity cannot – at this time – be solved. However, since both a magnetic field and the electromagnetic radiation can propagate through vacuums it would seem that something must be passing through these vacuums.}

{**Sided Comment on Dynamics and Kinematics**: The differences between kinematic and dynamic motion are examples of how one theory can antecede another. Kinematics is essentially a special (simplified) version of dynamics where the equations of motion are geometric trajectories and any forces that might be acting on the object of study are ignored, while in dynamics, the equations of motion include not just the geometric trajectories but include the forces acting on the object and the inertia (mass) of the object. Put simply, kinematics is about non-causal motion, while dynamics offers causal explanations for this motion.}

{**Side Comment on Scientific Definitions:** Science is a system based on definitions. That is, all propositions of a scientific theory are based on terms, which have been defined by other terms. At some point, there must be unproven terms such as axioms, definitions, and propositions that are accepted without proof, since this is the only way to avoid either an unending series of definitions or circular definitions. This essentially means that not all propositions within a scientific system can be proven, and this is one reason that theories are always subject to replacement or revision. Moreover, scientific endeavors are restricted to the unique set of notions that govern activities within each discipline (physics, chemistry, biology, ...). In this way, each discipline is a closed system within which the scientist must work. For example, a chemist *sees* the structure of matter differently from that of a physicist.}

{**Side Comment on Coordinate systems:** Science and engineering have long relied on the coordinate system devised about 400 years ago by the philosopher and mathematician Rene Descartes (French, 1596–1650). In two-dimensions, this is the familiar system based on two perpendicular lines generally referred to as the x-axis and the y-axis. Although this technique of using these XY-coordinates for science and engineering has been highly effective and useful, it represents a rather confined way of thinking. The problem has always been how to replace this system and with what?

The answer to this may be both somewhat old and somewhat new. Around the middle of the 19th century, an English mathematician, William Clifford (1845-1879), came up with an idea for a new algebra, but unfortunately, Clifford died before he could fully develop his new algebra. Consequently, his idea remained essentially dormant for about a hundred years until a physicist named David Hestenes (American, 1933-) rediscovered Clifford's work and pioneered a new geometry called Geometric Algebra, which has its beginnings with Clifford's algebra. This Geometric Algebra is essentially coordinate system independent, which makes it adaptable to many types of problems. Although this algebra has not replaced the old-fashioned Cartesian coordinate system, it is making gains and appears to have a bright future. By the way, you know all those dazzling special effects that you see in the cinema, geometric algebra plays a key role in making that possible.}

Scientific Revolutions of the 20th Century

In the first three decades of the 20th century, two revolutionary events occurred in science. The first was the development of the theories of relativity that forced a complete reevaluation of our classical notions of space and time. The second was the development of quantum theory, which was even more revolutionary than relativity. Quantum theory, as it developed, revealed that the behavior of matter at microscopic levels was stranger than could have been imagined.

The first revolution started in 1905 when Albert Einstein (German Physicist, 1879-1955) published a paper that eventually led to the *special theory of relativity*. At first, this paper did not receive much attention, and he did not become an overnight sensation. However, after several influential scientists read, and understood his paper, its significance began to be recognized. Even so, it was several years before his theories gained wide acceptance and brought worldwide fame. Old ideas are hard to give up and absolute time and space proved no exception.[26]

One of the few who, early on, recognized the significance of relativity was the mathematician Hermann Minkowski (Polish-German, 1864-1909), who, in 1908, made a speech where he pronounced that relativity theory connected space and time such that he coined the term *spacetime* (or *space-*

---

[26] It is worth noting that in 1905, not only did Einstein publish the paper on special relativity, but he also published three other papers, any one of which would have made his reputation as a first-class physicist. Moreover, one of these was a seminal paper for the future quantum theory and the one that later earned him a Nobel Prize.

time) to indicate this bonding.[27] That is, he proposed that time and space should be considered inextricably connected both mathematically and physically. In addition, to this paradigm change for time and space, the special theory introduced a famous equation that proclaimed the equivalence and mutual mutability of matter and energy. Several decades later, in 1945, the reality of this equivalence was dramatically demonstrated with the explosion of the first atomic bombs.

Ten years later, in 1915, Einstein, being no one-hit wonder, was again able to astonish the scientific world with the publication of his generalization of the special theory of relativity. As special relativity changed the status of time, general relativity changed the status of space. Put simply, special relativity specifies the speed of light as a fundamental constant of nature and demoted time to a variable (more about this later), while, on the other hand, general relativity specifies gravity as fundamental and space became subordinate to gravity. That is, within the *common interpretation* of general relativity, gravity determines what kind of space exists. Instead of the flat Euclidean space, space is assumed to be curved under the influence of gravity – later we will reconsider this notion about space.[28]

During the two decades that preceded 1905, the work of several scientists produced the building blocks that were later used by Einstein to create his theories. His intuitive grasp of the concepts and ideas enabled him to make new insights into the nature of things.

---

[27] Minkowski announced this proposal in a lecture delivered before the Congress of Natural Philosophers at Cologne (21 September 1908). Unfortunately, Minkowski died just months later and was not able to continue the development of his idea.

[28] Euclidean space is named after the fourth century BC Greek mathematician, *Euclid of Alexandria*, whose work called *The Elements* has essentially been in print for more than 2000 years! Euclidean geometry of space is analogous to flat surface such as a basketball court.

Others, especially Henri Poincare (French Scientist/Mathematician, 1854 – 1912), were close, but were never able to make the final synthesis. That is, prior to Einstein's 1905 publications, there existed a well-known principle called the *law of relativity*. Here is a definition of the law of relativity as given by Henri Poincare about 1898:

"The state of the bodies and their mutual distances at any given moment, as well as the velocities with which those distances are changing at that moment, will depend only on the state of those bodies, on their mutual distances at the initial moment, and on the velocities with which those distances were changing at the initial moment."[29]

Notice that here the word *absolute* is not used, and I might add that the *initial moment* is the start of measurement, not the so-called *big bang*. Clearly the notion of relative motion was well known prior to Einstein's papers of 1905, and the above-mentioned Henri Poincare, in particular, was actually very close to extending the law of relativity to the special theory of relativity.

Although, prior to 1905, some had considered the question of whether time was absolute or relative; it is special relativity, which has withstood all tests for more than one hundred years, that moved time from its privileged position as an absolute and immutable physical reality to a position of dependence as a variable. That is, time was no longer immutable but essentially depended on one's point of view.

{**Side Comment on Technique:** It has been suggested that Poincare, working as a mathematician, took a more formal approach; while Einstein, working as a physicist, used a less formal approach and relied on 'thought experiments.' Einstein was especially adept at this practice, which he used to help gain insight into physical phenomena in new and

---

[29] Poincare, Henri, Science and Hypothesis: New York: Dover 1952, page 78.

novel ways. For example, one of his thought experiments concerned what it would be like to travel at the speed of light.}

{**Side Comment on Relativity:** Put simply, the basic difference between special relativity and general relativity is the introduction of gravity, which is missing from the special theory. In special relativity, forces that move objects and the momentum of objects are not included, and this is one of the motivations for developing the general theory, which included both and more. General theory also includes gravitational stress and tension with the result that the mathematics of general theory are considerably more complex than that found in special theory. This is one reason why it took Einstein so long to develop the general theory, he had to learn new mathematics.}

Although the theories of relativity introduced radical changes in our understanding of space and time, it still maintained many of the characteristics of classical science. For example, within both relativity and classical science, nature is considered deterministic, causal, mathematically time reversible, and continuous. By retaining these classical notions, relativity may be considered the end product of classical physics, while quantum physics is the beginning of a new physics that no one had imagined possible.

*   *   *

Quantum theory was stranger than imagined possible because it indicated that at the micro-level, far removed from direct human experience, the behavior of matter was just the opposite from classical science. That is, quantum theory demonstrated that matter at this level is indeterminate, acausal, discrete, and, mathematically, time irreversible! Moreover, it seems that quantum theory has a mental element never encountered in science. While classical science used terms such as velocity, force, and momentum, quantum theory introduced into the lexicon of science unfamiliar and sometimes vaguely defined terms such as, uncertainty, entanglement, complementarity, duality, and exclusion.

Put simply, a large part of classical physics is about the behavior of matter at the macro-level, while quantum theory is about discrete states of matter at the micro-level (atomic or sub-atomic level). That is, classical mechanics is about smoothly connected behavior of extended matter found in both kinematics and dynamics. This is why differential calculus, which presents the appearance of unbroken connectedness, is so useful in classical science. On the other hand, in quantum theory, things do not flow; things apparently change states incrementally, without passing through any intermediate states.[30]

For example, the classical description of a rock thrown through a glass window is an event in which the rock has a specific trajectory that passes through the window and results in a broken window. On the other hand, the quantum version would be a description of the state of the rock and the state of the window prior to the window being broken and the state of the rock and the state of the window after the window is broken. There is no specific trajectory and there is no direct evidence that the rock broke the window, there is only the implication or probability.

Classical events have direct causality, while quantum events appear to have no direct causality. There are only before and after conditions or states. That is, there is no proof that the rock broke the window; there exists only a probability that the rock broke the window. That is, as stated by Gary Zukav, there is a *tendency* for certain things to happen and exist, which are expressed as probabilities.[31]

The replacement of the classical scientific notion of cause and effect with statistical *correspondence* was a necessary

---

[30] Physics uses the term *state* to describe possible results or configurations. For example, a coin toss results in either of two states – heads or tails, while the toss of a die has six possible states.

[31] Gary Zukav, The Dancing Wu Li Masters, (New York, NY, Perennial Classics, 2001), 35.

adjunct to the study of micro-phenomena. The gap between observer-interpreter (a human) and the microscopic physical system being examined is mediated through many levels of extrapolation, amplification, and simulation such that the results are not only statistical, but sometimes highly speculative.

The uncertain nature of quantum theory requires the introduction of mental elements as a means of bridging the gaps in the quantum level behavior of matter. Quantum events have a three-part aspect consisting of mathematical description, physical description, and mental description or interpretation. The mathematical (behavioral description) are in the equations of the theory. The physical descriptions are in the experimental setup and apparatus, while the mental elements are in the choices of what to measure and the interpretation of the instrument readings.

For example, the experimenter must make conscious choices about what will be measured. These choices include such things as measurements of either position or momentum - but not both, and wave-like or particle-like behavior - but not both. In this sense, the experimenter is no longer merely the *observer* of classical science but has become a participator.

Despite all the strangeness or, maybe, because of all the strangeness, the founders of quantum theory, for the most part, were not prone to examine or ask fundamental questions about the theory. It seems that this attitude was mainly due to Niels Bohr (Danish physicist, 1885-1962), who, as default leader, shunned fundamental thinking about the theory. Bohr held that quantum theory was a *statistically driven operational theory*[32] and that there was no good reason to seek a more fundamental understanding. With some exceptions, this attitude has continued to the present.

---

[32] Operational, in this sense, is like following a recipe which provides instructions without explanations.

As quantum theory developed, it became obvious that there comes a point in the study and contemplation of the natural world where the classical notions of matter, spacetime, and causality no longer applied, and a new (or perhaps very old) way of thinking is necessary. In between the determinism of classical science and the apparent randomness of quantum phenomena, lies the unpredictable and quasi-determinant middle ground where the randomness gradually gives way to determinism. That is, phenomena such as the famous *complementarity*, which is related to wave-particle duality slowly gives way to behavior that begins to resemble the predictability of classical science (See the *Side Comment on Complementarity* below).

There is even some speculation that this strange quantum behavior never completely goes away, but simply the probability diminishes asymptotically – in the sense of approaching but never quite reaching zero – as matter scales from the micro-levels to the macro-levels. This would mean that, although the probability would be extremely low, something strange like entanglement (See the *Side Comment on Entanglement* below) could occur at the level of human experience. For example, sudden and unexplained incidents might occur, such as those associated with synchronicity.

Exactly where this transition to diminished probability takes place is something of an open question. The wave-particle duality of quantum behavior has been observed in molecules with up to 60 carbon atoms, which is immensely greater in mass than a photon[33] or an electron,[34] but is still far below the macro-level of everyday experience. When physicists speak of the wave-particle duality, it should not be considered that observation of this duality means that the electrons or photons are actual particles or waves. That is, electrons and photons are neither waves nor particles, they are something else that only seems to behave like either a wave or a particle – depending on the experimental setup.

It is interesting to note that in 1906, JJ Thomson won a Nobel Prize by demonstrating that electrons are particles, and in 1937, his son, George Thomson won a Nobel Prize by demonstrating that the electron was a wave. Looks like you can have it both ways or either way – so, take your pick.

When thinking about *waves*, we tend to visualize them as having crests and troughs such as those observed on the surface of a body of water. This is OK, up to a point, but when contemplating the dual nature of particles, this description is totally inadequate. Wave-like behavior does not mean, in this case, an actually wave. This oscillatory behavior of particle is a mathematical description based on observed behavior, not necessarily a physical behavior. What it really is, we do not know, we only know that these particles exhibit properties that are inconsistent with solid matter, as we normally understand it (A conjecture about "What is light?" will be discussed later).

---

[33] Photon is the name given to the conjectured propagators of light, also known as electromagnetic radiation. These things, conjectured to be something like miniscule blobs of energy, only travel at the speed of light, and essentially do not exist other than when they are traveling at the speed of light.

[34] Electrons are elementary particles and are one of the basic building blocks of atoms. They are known to absorb and eject photons, as well as generate an electro-static field when motionless or an electromagnetic field when in motion.

{**Side Comment on *Seeing the Light*:** In his book on quantum electro-dynamics Richard Feynman insists that photons are particles.[35] This is in spite of the well-known wave-like behavior of photons as experienced in some experimental setups. However, given the discrete nature that is postulated for matter at the quantum level and the fundamental equivalence of energy and matter, it is reasonable to subscribe to the belief in the particle-like (or packet-like) nature of light (photons).

This, however, does not mean that light propagators are something like a grain of sand or speck of dust, which has definite size and shape. Photons and other micro-level particles have a different nature that allows them to manifest both wave-like and particle-like properties while being neither. We call them particles because we have no better word available.[36]

On the other hand, Feynman was incorrect when he said that quantum electrodynamic explains all of chemistry. For example, it explains neither chirality in nature (such as the fact that nature prefers left-handed molecules) nor does it explain emergent chemical properties.}

{**Side Comment on Quantum Jumps and Quantum Leaps:** One of the most common misconceptions and misuses of quantum theory are references to jumps and leaps. Scientists sometimes use the term quantum jump (or leap) when referring informally to the transfer, within an atom, of an electron from one energy level to another. Unfortunately, one often hears the expressions *'quantum leap'* and *'quantum jump'* used in the sense of a significant or notable change or enhancement to some thing or process. This is a strange and inverted usage of what is one of the smallest increment changes within nature, and it is very unlikely that anyone who actually knows something about quantum theory would ever use either expression in such a loose manner.}

{**Side Comment on wave-particle duality:** It seems that we attribute forces like electromagnetism to phantom-like elementary particles called photons. These *particles* far too small for direct observation, have an existence that is inferentially derived from the measurement of

---

[35] Feynman, 1985, 14, 37.

[36] At one time, the word *wavicle* was proposed, but it never caught on.

various phenomena arising out of experimental processes. However, these particles play tricks on us by changing their behavior. Perhaps the most famous and one of the simplest examples of this schizophrenic or bipolar behavior is the so-called *two-slit* experiment.

This experiment is conceptually quite simple, consisting only of a light source, an opaque screen, and a target such as a strip of film. First, two very narrow parallel slits are cut close together in the screen and then the screen is placed between the light source and the target. If one of the slits is covered and light is projected through the open slit onto the target, a solid bar of light appears on the target. However, if both slits are unblocked and the light is projected through both slits the expected result is two light bars each similar to the one observed when only one slit was open. However, instead of the expected pattern, a wave pattern appears on the target [alternating light and dark bars]. What this indicates is that, with only one slit open, the behavior of the light is particle-like, while with both slits open, the behavior is wave-like.

This means that you can measure either wave-like or particle-like behavior, but not both – at the same time. This behavior is known as complementarity. Do not try to figure it out; people have been trying, without success, for more than two hundred years! Thomas Young (English polymath, 1773–1829) first performed this experiment in 1803.}

{**Side Comment on Entanglement:** Entanglement is a condition where two (or more) particles such as electrons or photons become linked through some unknown process that binds them in a manner that is called entanglement. Metaphorically speaking, the result is that an action on one of these particles is instantaneously *felt* by the other. Essentially, the two particles act as if they were one particle. Additionally, this condition remains no matter how far apart the particles are separated. That is, spatial separation of any distance has no apparent effect on this condition. This phenomenon is another example of how quantum theory teases us by hinting at a reality that is yet to be discovered.}

{**Side Comment on What is Energy:** First, we know this is another one of those words that get used commonly in such expressions as

Today I feel low on energy or to remark that someone is *energetic*. This is fine but we also know that energy has something to do with fuel or caloric intake such as ingesting sugar. If we were to get somewhat technical, we would find that there are many different kinds of energy.

Here is a list of some common types of energy: kinetic energy, gravitational energy, electrical energy, chemical energy, radiant energy, nuclear energy, heat energy, elastic energy, and – last but not least – mass energy. Most of these are familiar, for example, batteries use chemical energy to produce electric energy, radiant energy from the sun fuels life on this planet, nuclear energy is used at nuclear power plants, and mass energy is the source of energy for nuclear bombs.

We know a lot about each of these different forms of energy, and they all provide capabilities that are essential for our world to function the way it does. Within science, technology, and engineering we use various mathematical formulas to calculate these numerical quantities we call energy, but after centuries of learning about these various kinds of energy we do not really know what it is – so I simply cannot answer the above question.}

Scientific Theories of Today

"...The research on the matching of general relativity with the quantum nature of matter is still in its infancy; however, we already understand that reconciling these two pillars of the XX's century physics will require a whole new set of ideas, as it may happen, that we may have to give up concepts that were supposed to be the starting point of all the modeling of the universe, such as that spacetime is a continuum and that the fundamental building blocks of matter are not point-like particles."[37]

"It is not always possible to check by experiment every separate assertion of a scientific theory, although the system of thought as a whole must, if it deserves the name of a scientific theory, contain possibilities of a check by empirical methods. This is what

---

[37] Orfeu Bertolam (2008). *The Mystical Formula and the Mystery of Khronos* (<arXiv: gr-qc/0801.3994v1>).

constitutes its verifiability. [Emphasis Added]"[38]

First, what is a *theory*? The dictionary uses words such as a supposition or a set of suppositions that have the intent to explain something. Other closely related words include hypothesis, conjecture, and speculation. However, instead of these words, the word theory is used a lot within the scientific community because it apparently seems more respectable and is considered to have a meaning closer to *fact* than the other words just listed.

Nevertheless, no matter how many times a theory is correct, there is always the possibility that it might require modification or even revocation. For example, the special theory of relativity has passed ever test of its validity for more than a hundred years, but is still ranked as a theory and has not been promoted to a *first principle* or *law*.[39] Put simply, a scientific theory includes these four things: a theoretical interpretation, a consistent mathematical formulation, the possibility of experimental confirmation (or the possibility of falsification), and prediction of previously unknown phenomena (that is, no hand fixing to match previously known observations). If any one of these things is missing, then it is a stretch to call something a scientific theory.

Put another way, to accept a theory is to take a speculative chance. There are always alternative choices, and the choice of selection is not always about better or worse theories, but rather about acceptability within certain established guidelines or intersubjective agreement based on personal and/or social agendas as well as the possibility of groupthink or other means of peer pressure.

There is a tradition that states, all things being equal, the simpler theory is, or should be, the theory of choice. In

---

[38] Wolfgang Pauli, *Writings of Physics and Philosophy*, Eds. C. Eng and K. von Meyenn. (Springer-Verlag: Berlin, 1994), 139.

[39] A *first principle* is a proposition that cannot be deduced from any other proposition, postulate, or conjecture. A scientific *law* is based on persistence of observations of the workings of the universe where no explanation is given, but there is an assumed causality.

practice, this is not always a correct method of selection. Simplicity alone does not suffice as a valid reason, there must be other considerations such as refutability, descriptive accuracy, evidential support, completeness, consistency, novel predictions, and so on.

For example, the theories of relativity are definitely more complex than the Newtonian theories that they superseded. Another example was the replacement of the geo-centric concept of the solar system with a helio-centric configuration that was not a simplification in that unlike the old model, where the earth was stationary, the new one requires that the earth move relative to the sun in both an annual orbit around the sun and a daily axial rotation.

Both of these examples profoundly changed the way we understand the natural world and obsoleted earlier theories. However, it seems that now there is a trend to patch-up theories, rather than replace the old or untenable theories with new theories. Additionally, not all of these patches are motivated by valid scientific reasoning but are made to save theories in support of non-scientific agendas.

Theories are not proven through repetitive observations or experiments; they are only made more probable. That is, inductive inference and consistency cannot prove a theory, it can only increase the probability of truth and increase confidence in the theory. On the other hand, one negative result can damage or even invalidate a theory.

It has been claimed that a necessary criterion of demarcation between science or some pseudo-science is its refutability. For example, according to Karl Popper, (Austrian British, 1902–1994), a system may be considered science only if it has been or can be tested by attempts to refute. As Popper said, "*A theory which is not refutable by any conceivable event is non-scientific. Irrefutability is not a virtue of a theory, but a vice.*"[40] Many consider this position too extreme, but, on the other hand it seems intuitively apparent that there needs to be some objective criteria for theory validation and acceptance.

Popper chose falsification for two main reasons, first, scientific theories such as Newton's *Laws* can never be fully verified but can be invalidated by a single false example. Secondly, by choosing falsification, Popper set up a demarcation between statements that can and those that cannot be found false. For example, in today's *Standard Model of Cosmology*, cosmic inflation is certainly not a falsifiable notion, and, in this case, it is also not a verifiable notion. As such, cosmic inflation should not be considered a true scientific theory but merely a speculative notion created to mend the big-bang conjecture [Discussed later in the Multiverse section - page 77].

True science is characterized by its observational and analytical methods, while pseudo-sciences are characterized by their speculative method (mental anticipations?) Irrefutability arguments may use either logical or empirical methods (or both?). An example of a *logical* irrefutability are the statements *'It is raining'* and *'It is not raining.'* These statements are logically irrefutable, although one or the other must be false. An example of *empirical* irrefutability is a statement of existential fact, such as *'I own a pair of brown shoes.'*

As another example, philosophical theories are irrefutable, but they may also be false. On the other hand, these days scientific theories fall into two broad categories; those that are empirical and those that are non-empirical in the sense that they are for the most part, if not entirely, observational (such as astronomy). Truly empirical theories are refutable, while the non-empirical are irrefutable for much the same reason that philosophical theories are irrefutable. Philosophical theories may be argued against, but falsification is not possible.

---

[40] Karl Popper, *Conjectures and Refutations: The Growth of Knowledge*, (New York, Harper and Row, 1963), 36.

That is, they are irrefutable in the sense of not being able to prove wrong, but, of course, are still deniable.

* * *

Science is obviously a human endeavor, but less obvious are the limitations or phenomenal boundaries within which scientific efforts are pursued such that the scientist is restricted to phenomenal information only within these boundaries. That is, each scientific discipline has its own boundaries that define domains of validity and discourse. For example, a chemist has ways of dealing with chemical elements and the structure of molecules that differ from that of a physicist. This means that if the physics community claimed to have developed something they considered a valid theory of everything, it would not necessarily be a theory of everything for a chemist, and, therefore, not really a *theory of everything.*

Scientists must necessarily restrict their deliberations and investigations to only a simplified version of nature whose complexity is too great for total consideration by any means currently available. That is, scientific studies are by necessity restricted to that which can be measured physically and analyzed mathematically.

There is an element of arbitrariness in scientific investigations because the scientist must make a choice of what to examine and what to ignore such that the real world as examined by a scientist is always a selective simplified version of reality [the reader might recall the Oppenheimer quote on page xi]. This is not to say that this technique is wrong, quite the contrary, this technique has, over the years, has been very successful. In fact, it has been so successful that scientists began to believe that they had captured the real world behind the world of phenomenal reality.

Of course, all these scientific discoveries and investigations have ignored and continue to ignore the realms of value and purpose which lie outside of the scientific domain that is, realms such as the mystical, transcendent, and natural beauty lie outside of the realms of science and technology. Put simply, scientists are sanctioned only to deal with the metrical, the qualitative and the quantitative of the phenomenal world and are not necessarily equipped to deal with these other realities.

Being restricted to the metrical domain, many scientists – under the influence of declining religious convictions – have adopted materialism. That is, lacking purpose or value they embrace a mechanistic view of life that reduces living creatures to mere machines.

Because of their specialized training, scientists, for the most part, seem to be neglectful of values beyond their chosen profession such that they have not been able to satisfy the natural human striving for that which is greater, and have attempted to satisfy this hunger through social and political agendas. Since many scientists today profess to be atheists and deny the existence of a non-material realm, they seem to seek their religious conation through extreme and exaggerated social agendas such as climate change where they have adopted the *chicken little syndrome*.

True and honest science exists only where there is open-mindedness and truthfulness along with a willingness to consider alternatives. It is worth repeating that the accumulation of evidence does not prove a theory, it only makes it more plausible or more likely to be true. This is not to say that valid theories are falsehoods or make-believe, rather they are only simplifications and approximations of reality, and this is likely to always be the case.

As mentioned in the first quote at the start of this section, along with the notion of a theory of everything, there is the current inability to unify the general theory of relativity (effects of gravity) with quantum theory (micro-level behavior of matter) and is an example of how physics has almost come to a halt. Given the contradictory nature of relativity theory and quantum theory, it is not surprising that all attempts to join these two theories have failed and will likely continue to fail unless some novel approach to the solution is found, which will likely fall outside of traditional mechanistic thinking.

It seems that the glory days of science are over – at least temporarily – and except for life sciences pure science seems to have made little real progress since the end of World War II, which, more than any other thing changed scientific research from an essentially individual effort into BIG science such as that which produced the first atomic weapons.[41]

Individual research and the pursuit of pure science is a very difficult task that requires more than just intelligence, it also requires years of hard work and perseverance. The problem is that few seem willing to do this or are the remarkable scientific revolutions of the early 20th century just too difficult to improve on in any real way. Or perhaps science in general has grown too fragmented through the increasing necessity to specialize such that the big picture that led to the past great discoveries is just too difficult to wholly grasp?

It is interesting to note, that both relativity theory and quantum theory are more than one hundred years old but there is still no recent major advancement of either these theories or some new theory that might have a similar impact as these older ones. OK, in all fairness, it should be mentioned that the theories of Newton (published in 1687) lasted more than two hundred years before they were rendered incomplete by the theories of relativity.

It should be recalled that the theories of relativity, as originally published, were essentially the work of a single person – building on the previous work of others, and the building of quantum theory, primarily in the 1920s, was the work of a handful of inspired and dedicated scientists.

Whether this lack of new game changing theories is because of the lack of insight or whether the pursuit of science

---

[41] What many today call science is really the engineering of new technology and new applications of existing technology.

has fundamentally changed such that *seeing* beyond these theories is just too difficult. It is a fact, that massive projects, like the LHC [*See Side Comment below*] in Switzerland, have provided employment for literally thousands of scientists and engineers, has cost billions of dollars, and to what end? Perhaps big science or science by committee is not capable of the type of insight that occurred up to the middle of the 20th century.

In any case, it seems that many of the so-called advances that are touted in the press are not particularly reliable with regard to being really scientific or are they just pipedreams of wishful thinking masquerading as true science?

Perhaps physics took a wrong turn and became lost in theories and conjectures that have led nowhere? Some have even speculated that all attempts to merge relativity and quantum theory are doomed to failure because of incompleteness and/or inconsistencies in both theories. In any case, physics seems stymied and at a stalemate.

Here is a short list of some current issues:

* String theory and M-theory came and went.
* Quantum gravity seems to be going nowhere.
* Unverifiable conjectures such as the *multiverse* have gained in popularity.
* Particle physics seems like a self-fulfilling prophecy.
* Unquestioned adherence to the big bang and cosmic inflation conjectures.
* Unquestioned adherence to the Standard Model of Particle Physics.[42]

---

[42] Wikipedia definition of the *Standard Model of Particle Physics*:
"The Standard Model of particle physics is the theory describing three of the four known fundamental forces (the electromagnetic, weak, and strong interactions, but does not include the gravitational force) in the universe, as well as classifying all known elementary particles."

These are just a few of the many issues in physics and fields related to physics. If we were to venture into biology, chemistry, and so on, we would find that these fields also have many unsolved and, perhaps, unsolvable problems.[43]

* * *

The empirical knowledge of science is always subject to revision due to some future insight, measurement, or observation. Scientists deal with phenomena and must always be aware that beyond the phenomena is the unknowable thing-in-itself. Given this, it is not surprising that science uses conditional words such as theory, postulate, hypothesis, and conjecture when dealing with its subject matter. There is always an element of incompleteness and uncertainty.

The Standard Model of Particle Physics is a prime example of what has been sometimes called an infectious idea that grows in both content and complexity in pursuit of some final conjecture. This is why the number of so-called elementary particles exploded out of control and created a culture with an implicit particle mentality. That is, true believers apparently cannot conceive of any solution that is not based on elementary particles.

As stated previously, no current scientific theory will escape modification or possibly even revocation and replacement. All scientific knowledge is in flux and as we learn more about our place in the universe, even religion is not exempt from change and reinterpretation. In any case, up to the present, efforts to develop a theory of everything or a Grand Unified Theory, as it is sometimes called, have not been successful and these efforts seem likely to remain unsuccessful.

---

[43] For those who might be interested, Wikipedia has lists of unsolved problems for physics, chemistry, biology, and more.

{Side Comment on Particle Physics and the CERN LHC: CERN is the *European Center for Nuclear Research* located in Geneva, Switzerland. This facility includes the *Large Hadron Collider* (more commonly called the LHC), which is the world's largest and most powerful sub-atomic particle accelerator. The LHC is a massive group of machines, which includes a circular tunnel 27 kilometers in circumference. Its purpose is to investigate the nature of matter at microscopic levels, well below the size of atoms.

One of the main justifications for this multibillion-dollar operation was expressed as the search for a particle known as the Higgs, which is named after the physicist who first proposed the possibility of its existence. Anyway, about the time that the LHC went operational in 2008, there was no doubt in my mind that the Higgs particle would be found. This is not because I believed in the existence of the Higgs particle but because they <u>had</u> to find it. This was the only way they could justify the great effort and massive expenditure.

Sure enough, after a few years of operation, they eventually claimed to have discovered the Higgs, and this discovery was officially published in the 2012 edition of the Review of Particle Physics. This document contains 1526 pages and mentions Higgs, 1122 times, but what is more interesting is that the word assume (or variations on this word such as assumption) is used 522 times![44] Obviously, Higgs is the star of the show, but the use of so many assumptions seem to imply a lot of guesswork rather than hard evidence.

By the way, the 2016 version expanded to 1808 pages, mentions the Higgs 1823 times, and uses the word assume or one of its variations 723 times![45] In the most recent 2020 version (2092 pages) it seems that the Higgs is more popular and is mentioned 2272 times and the word

---

[44] The 2012 edition of Review of Particle Physics was published for the Particle Data Group in Phys. Rev. D86, 010001 (2012) [http://pdg.lbl.gov/]

[45] The 2016 edition of Review of Particle Physics was published for the Particle Data Group in Chinese Physics C Vol. 40, No. 10 (2016) 100001 [http://pdg.lbl.gov/]

assume or variations (assumption, assuming, etc.) increased dramatically to 4845 occurrences![46] The fun never stops!}

{**Side comment on Creativity:** Consider these three words; *discovery*, *invention*, and *creation*. They represent three different ways we advance knowledge and all three have very different meanings and reference very different activities. Discovery is what explorers and archeologists do. They find things and then try to attach some significance or meaning to what they find. On the other hand, inventions, which might be described as the assembly of some novel thing using known and/or unknown parts. This is what engineers, tinkerers and, possibly, some scientists do. At the peak of this hierarchy of these three words, is creation. True creation is, among other things, an act of seeing things differently. Creation is not about quality or even esthetics, but it is about uniqueness and not necessarily utility or application. Put briefly, creation is the act of producing something unique and is more than invention, as invention is more than discovery.}

Scientific Dogmatism

"The essence of scientific freedom is the right to come to conclusions which differ from those of the majority." [47]

At the beginning of modern science, in the West, science had to be careful about conforming to the dogma of the western church. Today, instead of religious dogma, a scientist is required to follow the prevailing scientific dogma. In the 16th century, at the beginning of modern science, one could lose one's life, today one can remain unemployed [See *Side*

---

[46] The 2020 edition of Review of Particle Physics was published for the Particle Data Group in Phys. Rev. 1:
https://pdg.lbl.gov/2020/html/computer_read.html. The 2023 report is contained in multiple PDF files so here is a link to the 2023 reports:
https://pdg.lbl.gov/2023/reviews/contents_sports.html.

[47] Edward Arthur Milne, *Modern Cosmology & the Christian Idea of God* (Edward Cadbury Lectures, 1950): (Oxford: Clarendon Press, 1951), 8.

Comment on Giordano Bruno – page 96]. The punishment is different, but the result is the same, innovation and freedom of thought are stifled.

It is the goal of all scientific endeavors to determine objective characteristics of the natural world. However, at the same time, science is also trying to undo, modify, or extend these previously determined objective characteristics. The result is at best a complementary situation, but, in practice, it is more like a conundrum or contradiction. There exists a sort of *push-pull* effort between the desire to discover the new and the desire to preserve the old. Failure to entertain novel ideas or the possibility of change leads to stagnation and dogmatism, while readiness to consider novel ideas leads to innovation and true progress. Of course, this readiness should not be hasty or premature acceptance of notions with little or no valid supporting evidence.

It seems that the practice of science has moved from an understanding of its limitations and what it actually states, to a more socially motivated position where the limitations of knowledge about the natural world are encased in elaborate theories that are proclaimed to the world as fact and truth about nature even though many of these statements rest on weak foundations or are simply made-up ad hoc statements used as patches to some grand conjecture.

For example, there has been a trend of making proclamations about the discovery of remote planetary systems which total in the hundreds. This trend is reminiscent of the proliferation of so-called elementary particles that grew with each increase in accelerator power and eventually numbered in the hundreds.

I do not doubt that billions of other planets are out there, I just recommend caution concerning these *discoveries*. These findings should be considered preliminary and not a proven scientific fact. Error rates have been estimated as high as 35%, which is not far from guess work.

Without doubt, reluctance to consider novel ideas has become dogmatically engrained within the scientific community. Those who dare question the orthodox risk being *burned at the stake* of rejection and ridicule! Unfortunately, this is not an entirely new thing, as far back as 1927, at a conference in Copenhagen, Louis de Broglie (1892 – 1987), an eminent French physicist, was treated rather badly by several of his colleagues. De Broglie had made a proposal that did not fit the current quantum theory dogma as dictated by the inner circle at Copenhagen. Only tears later were his ideas given a fair hearing.

The community of physics orthodoxy is full of mathematically competent practitioners who scorn and ostracize dissenters, such that there is no longer any place within this culture for free thinkers of the type that dominated physics until the middle of the 20th century. These innovative physicists of the first half of the last century set a very high bar for hard work, creativity, and accomplishment such that few if any remain with that level of work ethic and capability. Today, as a result of big science, today's technocrats bow down to half-baked conjectures that seem to pop up without justification other than the belief that they need to look like they are making progress.

* * *

As the scope of scientific knowledge grew, it began to split into ever more specific areas of specialization, which caused a fragmentation and isolation between these specialized areas. To be sure, the vast scope of scientific knowledge makes it impossible for any individual to grasp much more than their chosen discipline. Consequently, the age of the *polymath* or multi-disciplined natural scientist is essentially over. Complexity is reduced by dividing large problems into smaller chunks that can be manipulated and in this way problems that otherwise might prove too difficult can be successfully solved. The downside is that this results in even more fragmentation that makes it easy and even likely that one will lose sight of the connections that link these scientific divisions.[48] As a result of this increase in narrowness of focus, the need for scientific collaborations has increased greatly such that now there are consortiums that employee hundreds (or even thousands) of scientists and engineers in pursuit of solving problems and understanding nature.

What is Light?

"...one should not think simply in traditional forms of the explanation of nature, even if these forms are well formulated in historical philosophical systems but should be open-minded for new logical and empirical possibilities, not foreseen by these systems."[49]

The simple answer to this question is that we really do not know. On the other hand, we do know a lot about light behavior, and it is this behavior that provides clues about the true nature of light but, unfortunately, not enough to know light's true nature.

* * *

Before getting into the main topic, it seems appropriate to briefly discuss some of the known behaviors of light. Although most 'facts about light' make references to stuff like the velocity of light and usually include something about photons, here are a few different facts.

- Some frequencies light (electromagnetic field) can be harmful to humans and other living things.[50]

---

[48] Another part of the problem is the plethora of scientific papers that are being published. For example, astrophysics papers at ArXiv.org (Cornell University Library Archive) have several sub-categories that together result in about one thousand new papers each month!

[49] Pauli 1994, 139.

[50] The electromagnetic spectrum - wherein visible light is a part – ranges from AM radio which has a wavelength of around 100 meters to x-rays and

- Light, at certain frequencies, influences metals such as antennae.

- Various frequencies seem not to interfere with each other. The immediate space around us is full of electromagnetic radiation from a plethora of sources such as radio transmitters, microwave ovens, cell phone towers, and so on.

- Some frequencies can penetrate wood and other non-metallic objects, which is how we are able to use a radio inside a building without windows.

- It seems obvious that visible light is easy to block – for example, a thin piece of cardboard can block most sunlight. This is strange since visible light is about in the middle of the electromagnetic frequency spectrum which indicates more energy than radio signals which have lower frequencies and, on the other hand, x-rays and the like have far more energy than light signals. As is well known, both radio signals and x-ray signals can easily penetrate solid matter while visible light, which lies between the low energy radio signals and the high energy x-ray signals, cannot penetrate walls! So, what is going on? It is not surprising that more energetic signals can penetrate solid matter, but if visible light signals – unless amplified greatly – cannot penetrate solid matter then why are the less energetic radio signals able to do so?

- Light at any energy level or intensity cannot be seen in transit – when we see a beam of light, we are seeing the effect of the light on particles of matter suspended in the atmosphere. Only *intercepted* electromagnetic radiation is visible. Consequently, it might prove impossible to ever discern its true nature.

- Our knowledge of electromagnetic force is indirectly obtained and really does not explain how light can travel billions of years without being diffused, absorbed, or otherwise diminished to the point of being undetectable.

- Another measured fact is the independence of the speed of light from the relative movement of both its source and receptor. Although it is also true that light slows down while passing through

---

gamma rays which have wave lengths of about 0.1 nanometers while visible light is in the 400 to 700 nanometer range.

a transparent material then returns to its *normal* speed on exit from the transparent material.

- It should be noted that visible light (for humans) represents only a very small portion of the known electromagnetic spectrum.

* * *

Anyway, this discussion will deal partially with what we know about light behavior and offer a supposition (speculation) about light itself in a general sort of way – since that is about all that we can really do. That is, only hints or vague statements can be made about the true nature of light – even though there are those who mistakenly claim to know that true nature.

Regarding the behavior of light there is one thing that has been known for more than two-hundred years and for all that time has been a conundrum. This is the above mentioned *duality problem*, where depending on the fact that light sometimes behaves like a wave and at other times like a particle.[51] In fact, as I have mentioned elsewhere herein – page 23, the Nobel Prize in physics was awarded to one person for demonstrating the light consisted of particles, and some years later his son won for demonstrating the light was a wave [See page 22]. So, there you go, sometimes you can have it both ways.

Before getting into the subject itself, I should mention that the following discussion is tightly bound to ideas put forward by Alios Wenzl (German philosopher: 1885 – 1967) in his 1949 essay written for the *Library of Living Philosophers* volume that honored Albert Einstein [numbered quotes are from this essay].[52]

----

[51] Thomas Young (English polymath, 1773–1829) first performed the experiment that indicated this dual nature of light in 1803. This experiment is sometimes referred to as the 'two-slit' experiment.

[52] *Albert Einstein: Philosopher– Scientist*, The Library of Living Philosophers, Ed. Paul Arthur Schilpp. Tudor Publishing Company NY, NY (1949).

*   *   *

The Problem of Light: [partially quoting Wenzl] How can light from a moving source (for example: the sun) that spreads out [1] *"independently of the motion of the source – equally in all directions, if it is a [physical] reality – which moves, then it cannot proceed with objectively equal velocity with reference to an observer A, on the one hand, and with reference to an independent observer B* [Emphasis added]." To resolve this difficulty, Einstein proposed that the velocity of light was constant. That is, in contrast to the above intuitive statement about light as a *moving reality* is the constancy of the velocity of light postulate,[53] which is at the heart of Einstein's theories of relativity and has been criticized over the years as being arbitrarily determined. However, despite this criticism this notion has also been consistent with experience. That is, experimentally verified.

How is it possible when considering the preceding comments about the two independent observers (A, B), along with the constancy of the velocity of light postulate, and the duality problem while at the same time [2] *"avoid the contradiction that a physical reality [physical body], which moves, cannot move with objectively identical velocities with reference to different observers* [See quote #1]? *This indeed can be done, but one must draw the conclusion that light may neither be treated nor be viewed as a material body [particle] nor as a wave moving in three-dimensional space* [Emphasis added]."

Before continuing, it should be remembered that, today, most scientists consider light to be composed of microscopic particles called photons, which means that if light were particles, these particles would necessarily be finite in number such that during transit they would separate according to an *inverse square rule* and would therefore become extremely diluted. For example, if stellar light is in transit for millions or

---

[53] This postulate states that light velocity does not change due to relative motion between the light source and the light recipient.

even billions of years – as is now declared by astronomers, astrophysicists, and so on – this dilution seems likely to make any detection of these remote sources extremely difficult or even impossible.

However, since we apparently can detect light that has been in transit over such an immense number of years, it seems a reasonable and even necessary conclusion that- [3] *"we dare not regard light as moving particles nor as real waves of a system-bound medium. On the other hand, there is no reason why light should not be regarded as a signal which actually behaves toward all systems disinterestedly and impartially, which is neither imprisoned in nor bound to a material system but is super-systemic (or above all systems). Only in that case we may not regard it [light] as something already material but must consider it as something 'immaterial,' 'pre-material,' [or] 'potential'* [Emphasis added]."

The notion of *potential* has been around at least as far back as Aristotle and, today, is very much a part of modern physics where, for example, potential energy is regarded as a basic concept. This raises the question: [4] *"What does it mean to say that light is above all systems?"*

One answer is that light can be regarded as [5] "the signal of a change in existential relations, no longer and not yet materialized and therefore not yet system bound, but rather, as the dissolution of a former material order, the mere announcement and communication of a new possibility. It is as yet a potential situation, which does not refer to a single system, but which restores the disturbed order precisely because of its reference to all [systems]." Put another way, light – in transit – can be considered [6] "as a field of possibilities to which all systems belong," and it is this field of potential which successively actualizes according to the relative distances of recipients from the signal's origin. It is this actualization that we refer to as photons or in Wenzl's words – [7] "potential energy propagates in order to actualize itself again when it hits a wall. "

What moves ahead is therefore the super-systemic potential such that in every system the changed existential situation can [8] *"announce itself successively according to the distance relation between the origin of the disturbance and the recipient."* In this manner the duality of light is resolved such that the wave nature corresponds to its potential state and the particle state with its actualization in a system. [9] *"If, therefore, one speaks of the undulatory or corpuscular nature of light, the contradiction, which has created so many difficulties, is solved."*

*   *   *

As noted at the beginning of this section, Wenzl's essay dates from 1949 and I suspect was soon forgotten. It is obviously a stretch of the imagination, and such a notion would be difficult to verify since the only way to experiment with light is to have a recipient intercept the light which would – according to this speculation – change its nature through actualization.

To be sure, I promised something vague, and the term *super-systemic* certainly meets that goal. Nevertheless, since this notion is based on known behaviors of light, I do believe that there may be something worth considering in this notion which seems more reasonable than some of the other ideas such as the *cosmic inflation* and *multiverse* fantasies that are without any physical or experimental evidence, but have been put forward as being scientific when, in fact, they are just made-up stories.

Reconsidering Time and Space

History has shown that those who question the orthodox are usually the innovators. Giving up long held beliefs is never easy, but often necessary. We need to remember that stepping out of our conventional beliefs, like stepping out of the door, can lead to the unexpected. Just ask Alice...or Bilbo...

Time Reconsidered

"It was shown already by Aristotle ... that the psyche appears to be responsible for the 'measuring' and 'the counting' of (physical) motions and thus, for the coming into being and the genesis of chronos at the human level as a psychological phenomenon with ontological pretension. "[54]

"How may it then be measured? And yet we measure times; but yet neither those which are not yet, nor those which no longer are, nor those which are not lengthened out by some pause, nor those which have no bounds. We measure neither times to come, nor past, nor present, nor passing; and yet we do measure times." [55]

As previously discussed, we measure common time by observing the change in position of an object, such as the *hands* of a clock. All objective measurements of time require physical motion, whether it is the daily passage of the sun or sand falling within a glass enclosure, the measurement of time has always been through some cyclical *physical* process, while our understanding of time, through experience of its variability, is a *psychological* process.

In this section, we dig deeper into the concept of time to determine its true nature and to do this it will be necessary to let go of the notion of common time and open ourselves to novel ideas, and like all well entrenched and seemingly intuitive concepts, letting go of common time may not be easy.

---

[54] Attila Grandpierre, "The Dynamics of Time and Timelessness," *The Nature of Time: Geometry, Physics, and Perception.* NATO Science Series II. Mathematics, Physics, and Chemistry - Vol. 95, (Netherlands: Kluwer Academic Publishers, 2003).

[55] St. Augustine of Hippo, 129.

Order

Before continuing with time, it will be useful to have a brief discussion about the word *order*. Again, like time, here is a common word used without much thought about what it really means. Although order has many usage definitions,[56] the sentential context usually provides enough clues to determine the intended meaning. Even though there are many types of order, in its simplest form, we tend to think that order has something to do with the arrangement of physical objects. For example, objects in a straight line are said to have *order*.

Another example is the sequential ordering of events, which, it seems, is an essential element of our understanding of common time. We build up our common notion of time from sequences of events, which we perceive as having a sequentially relational order. These events appear to occur in a continuous sequence that leads us to think of time as having a certain continuity. Sometimes, we call this sequential order the flow of time and, sometimes, this flow is associated metaphorically with flowing water.

It seems that a natural tendency for humans is to seek order and even go as far as imagining order where little or none exists. But to say that none exists is a human reaction to what may just be too complex for understanding. That is, is what we call chaos really chaotic or just beyond our comprehension of what we call order. For example, is the randomness attributed to quantum theory really random? Put another way, due to our sensory and cognitive limitations, we are only able to discern order in relatively simple situations such that as complexity increases, we eventually lose our ability to see order and so we call it random or chaotic.

In a very real sense, the word *order* defies an exact definition because it has a vast scope of usage and through

---

[56] The online Oxford English Dictionary lists about 50 usage definitions for the word order.

contextual meaning the word *order* encompasses many aspects of human experience. Everything, from the simplistic process of lining things in a row to the complexity of the natural world, such as worldwide climate change and human society, has some level of order even though in the more complex situations the order may be difficult to detect.

Anyway, here is my questionable stab at a generalized definition. Order is giving attention to similar-differences and different-similarities. That is, order is a mental condition based on comparisons such as ratios, proportionality, and so on. Our notions of order clearly depend on an ability to perceive similarities and differences, which implies that, by extension, order is connected to our notions of categories and concepts.

For example, an important aspect of order is the ability to measure, and all measurements are a comparison which implies proportionality. Also, there is the comfort factor in measurement such as instead of the subjective statement of "I believe I have a fever," a measurement with a thermometer adds credibility to the notion of having a fever.

Our conceptual understanding of order is not intuitive but is one of the first things we learn. Starting in early childhood, parents and others teach order through the arrangement of things. First, just by lining things up and later as we develop language skills, our sense of ordering becomes more complex and terms like *alphabetical order* and *numeric order* become familiar, which, in turn, leads to more mature notions about order.

When establishing order, whether of events or objects, we learn to understand and use relational concepts such as *before, after,* and *in between.* The first two are binary relations between two objects or events and the third is a tertiary relation between three objects or events. Put simply, if event A is *before* event B, and event C is *after* B, then B is *in between* A and C. These learned notions are so familiar that we rarely give them any thought and are examples of conceptual thought. To be sure, these notions about order are well-known and learned at an early age, nevertheless, something is still missing.

We may know the sequence of events, but we really do not know anything definite about the intervals or gaps, which separate these three events (A, B, C) such as - are the intervals between these three events equal? We do not really know unless we have some standardized or common means of measuring these intervals. Just as we might use a ruler, which is a standard linear unit of measurement, to measure an object's extension or the separation between objects, we need something equivalent to the ruler that will measure the intervals between events as well as the duration of events.

What Does a Clock Measure?

"The concept of a clock enfolds all succession in time. In the concept, the sixth hour is not earlier than the seventh or eighth, although the clock never strikes the hour save when the concept bideth."[57]

That may seem like a silly question since *everyone knows* what a clock does. It tells time. Of course, that is wrong; we only assume that it tells time. What it really does is something else, but before answering the above question, we first need to think about what a clock really is and what it actually does.

Certainly, for everyday common usage, the definition of a clock as given, for example, by the online *Oxford English Dictionary* as "*a mechanical or electrical device for measuring*

---

[57] Nicholas of Cusa (1401-1464), *Vision of God*, 1453, 24.

time" is completely adequate.[58] It only becomes inadequate when we delve deeper into the nature of clocks and time.

All clocks are devices that repeat some type of cyclical *motion*. Furthermore, we assume that this cyclical motion repeats at equal intervals. There is just no other way of measuring event intervals except with predictable repetitive motion associated with an object that is external to the object of measurement. That is, if I want to measure a series of events, I necessarily have to use a measuring device that is independent of the events being measured.

Clocks, in a monotonous manner, simply repeat mechanical processes with what we faithfully believe are equal intervals. These repetitive processes require an *interpretative interface* through which the repetitive motion acquires meaning. As Nicholas of Cusa wisely observed six hundred years ago, six o'clock is not a characteristic of a clock; but has only a symbolically interpreted meaning that we give to a particular configuration of the clock-human interface.

The human interface of a clock is only a visual mechanism of successive positions and is indifferent to the hour. Although we commonly assume that a clock is *telling time*, what clocks really do is to provide a standard measure against which we determine, through our interpretation of the clock-human interface, the intervals between events and the duration of events.

The history of clocks is rich and varied, and today we encounter clocks in many configurations. Nevertheless, all clocks still have essentially the same purpose, and they all rely on processes that meet our expectation of reliably repeating a *cyclical motion with equal intervals*. That is, they rely on the repetition of a particular motion with intervals that are assumed to be equal within some acceptable degree of

---

[58] To be sure sundials have no mechanical moving parts, it this case, it is the earth's movement that provides the measure of time.

accuracy. This is what all clocks do, whether it is sand falling through a glass tube or a modern sophisticated *electronic clock.* [59]

Does Time Flow?

In the world of everyday experience, time, whether continuous or discrete, certainly seems to flow in a continuous manner and is sometimes compared analogously to the flow of water in a stream. That is, water in a stream appears to flow in an unbroken manner, but as we are aware, the water in a stream only seems continuous. Actually, the stream is an aggregate of an extremely large number of water molecules and therefore, at the micro-level, this flow is discrete and not continuously connected.[60]

Direction of Time

The idea of the flow of time is still not complete. The flow needs direction. At this point, you might be tempted to say that this is obvious – time flows from the past to the future. Well, for all practical purposes, this notion of the flow of time seems correct, but as with other common usage, this statement needs closer examination.

---

[59] Even electronic clocks use cyclical or oscillating motion as the measure of interval. There is also the so-called atomic clock that uses the International System of Units that defines the second as the duration of 9 192 631 770 cycles of radiation corresponding to the transition between two energy levels of the caesium-133 atom. This is still a mechanical process.

[60] An 8-ounce (224 grams) glass of water has about 7.5 X 10^{24} molecules of water. (10^{24} = 1,000,000,000,000,000,000,000,000). To put this into perspective, it has been calculated that there are more molecules in an 8-ounce glass of water than there are glasses of water in all the oceans!

If I move the cup on my desk from one position to another, I can always move it back to the original position. In doing so, did I reverse time, or did I simply move an object without any influence on time? All we can say about the two moves is that we only changed the position of the cup. We did not change its appearance or *state*. That is, these moves left the cup itself unchanged and after the second move, it was as if the cup never moved. Both cup moves are examples of *reversible* processes and will not help us determine the direction of time.

Instead of just moving the cup, what if I drop the cup to the floor and the cup breaks. We can all agree that the cup is not likely to jump back onto the table and become a whole cup again. This is a very different situation from just moving the cup from one position to another and back again. The breaking of the cup is an example of an *irreversible* process. Since the cup went from an unbroken state to a broken state, it may well be that irreversible processes might be useful in determining the direction of time.[61] Another rather common example of an irreversible process is simply eating a sandwich.

Past, Present, Future

"Predictions are difficult, especially about the future."[62]

In everyday experience, we often encounter the terms past, present (or now), and future. And once again, just as with the common words previously discussed, we use these words with little thought or understanding of what they really mean.

---

[61] The direction of time is sometimes called the arrow of time. For a good discussion of various theories on the arrow of time I recommend the Stanford Encyclopedia of Philosophy, http://plato.stanford.edu/.

[62] Quote attributed to Niels Bohr, a Danish physicist and one of the founders of quantum theory.

The future: Of these three words, the word future is the least problematic. We anticipate the future based on our experience of past events, our expectations of future events, and our wishes (anticipated desires). Future events are undecided and have probabilities of occurrence running from extremely high to extremely low. For example, the certainty that the sun will reappear in the morning and on schedule, to the latter which includes low probability events such as winning a lottery.

In the middle, somewhere are undecided and uncertain events where with reasonable degrees of probability we anticipate and expect a resolution. This includes events such as athletic competitions, where an outcome is certain, but the exact outcome is speculative. Here we have an example of a subtle and important distinction regarding the status of certain future events. An event that is determinate – it will be settled, but not pre-determinable – results not known in advance.

The future, then, is the realm of possible events, which we look forward to with anticipation, anxiety, and sometime fear. The future is unrealized potential.

The past: Events of the past are fixed in the sense that they cannot be altered, and, for most of us, events of the recent past are not problematic in the sense that we are capable of recalling actions and events of the recent past with an acceptable degree of accuracy. For example, finding an object today that was placed somewhere yesterday is usually quite easy. The past, when it concerns recent events, is usually neat and tidy.

On the other hand, the reliability of memory for events in the more distant past can be problematic. It is an unfortunate and inconvenient truth that memory – like the oral traditions of ancient times – is not very reliable. This is because, either consciously or subconsciously, we regularly alter our memory of past events.[63] For example, we often modify unpleasant and emotional memories to make them more acceptable and tolerable. Of course, sometimes, we just forget.

Mostly, we base our future actions on remembered experiences based on often modified or prejudiced recollections. That is, our prejudices - for indeed we all have them - unavoidably influence our actions. The past is fixed, but recollections of the past are not and, besides, no two individuals share the same past.

The present: Briefly, the present is the fleeting middle ground that is wedged between the past and the future, which on recognition as the present has already become the past. Since the *present* is where we seem to spend our lives, it is somewhat more complex and elusive than the future or past, so we will need to dig deeper to see what is really going on.

What now?

The truth of the matter is that everything we experience has already happened. There always exists a delay in the reception of sensory sensations and an additional delay in the comprehension of this data. For example, whether the perceived event is a visual image, a sound, or a combination of both, these events have a short but real delay due to the finite speed of light and the much slower speed of sound.

Additionally, there is the delay of comprehension that must follow the reception of the sense data. In this way, the *now* has already slipped into the recent past before we are completely aware of it. Of course, this happens very quickly so

---

[63] I have purposely used the word subconscious because the word unconscious is often associated with not being awake or lack of awareness. Unlike Freud, I have no problem understanding that subconscious does not refer to spatial sub-structure but to a hidden part of the Mind.

that the delay is subtle to the point of being unnoticed and most events give the impression of being immediate. This essentially means that all references to the present (or now) are always somewhat vague and that such references are only appropriate for informal or common discourse.

However, the notion of the present is deeply ingrained in the human psyche, and all efforts to change our ideas about the present are, at best, difficult or possibly even futile. Perhaps, the best we can do instead of thinking in terms of some instantaneous now or present, is to modify our idea of the present and adopt an extended version of the present. That is, to smear it out – like passing a finger over ink not yet dried.

Duration

"Duration can be all nature present as the immediate fact posited by sense-awareness."[64]

"I will use the term 'moment ' to mean 'all nature at an instant.' A moment, in the sense in which the term is here used, has no temporal extension, and is in this respect to be contrasted with a duration, which has such extension. What is directly yielded to our knowledge by sense-awareness is a duration."[65]

"Duration is the continuous progress of the past which gnaws into the future and which swells as it advances." [66]

In the course of our daily lives, it seems apparent and intuitive that events flow continuously. That is, it appears that each moment connects to the next without a gap so that the movement of an object appears to be continuous. Thus, we perceptually conceive that the passage of events through time

---

[64] Alfred North Whitehead, *The Concept of Nature*. (Cambridge: Cambridge University Press, 1920), 24.

[65] Ibid, 57.

[66] Bergson 1911, 5.

and space is unbroken. However, this apparently continuous flow is an illusion. What we really experience is a sequence of images or durations that we mentally interpret as being a continuous flow.

That is, instead of a continuous flow, we experience a sequence of discrete durations where the current duration is sandwiched in or delineated by the beginning of the next duration and the end of the prior duration. This ensures sequential continuity between previous durations and future durations.

Perceptually, the flow of these durations smears out, or runs together, with no definite beginning or end. These sequences of smeared out durations comprise what has been called the *specious present* and through this continual sequence of durations we experience both the flow and direction of common time.

Specious Present

In 1892, the eminent American psychologist and philosopher, William James (1842-1910) was able, using some existing and some novel ideas, to redefine the present in a form that he called the *specious present*.[67] He defined the specious present as, "a sort of saddle-back of time with a certain length of its own, on which we sit perched, and from which we look in two directions into time." Although some have questioned the use of the word specious, it seems appropriate since specious is defined as "superficially plausible, but actually wrong or fallacious."

That is, the specious present is the *window* through which we *observe*, both forward and backward, the sequence and succession of durations. This is similar to the way we

---

[67] William James, *Principles of Psychology the Briefer Course* - 1892 (Cambridge MA: Harvard University Press, 1923), 280.

interpret musical notes, not as individual tones, but as a flowing sequence of notes, which seem continuous rather than individual. Additionally, in the case of familiar music, the anticipation of future notes also contributes to the apparent continuous flow. Essentially, we trick ourselves into thinking of music as a continuous flow.

Some say that succession through a temporal sequence of events relies on the permanence of objects. That is, some degree of permanence is necessary for knowledge of the passage of time, while at the same time, perceived change is the only way of measuring this passage. At first this may seem like a contradiction until it is recalled that permanence does not preclude cyclical changes such as those that occur regularly within the natural world. Of course, perceived change is not just physical but includes the succession of mental durations (sequence of thought).

The same holds for any sequence of events, which occur at a rapid enough pace to seem continuous, just as the pace of each frame of a video appears in our consciousness as an unbroken flow. Both the apparent flow of music and the continuous flow of a video are dependent on a certain amount of perceptual persistence and on our minimum threshold of perception.

The specious present is concomitant with the flow of durations, which results in perceptual continuity sometimes called the passage of time, but more correctly, should be called the passage of nature. The passage of nature can be as short and fleeting as a flash of lightning or more drawn out such as the growth of a garden. The passage of nature is our awareness of the world moving through our consciousness.

Tensed or Tenseless Time?

In 1908, J. E. McTaggart (English philosopher, 1866-1925) wrote an article about time for the quarterly journal *Mind*,[68] in which he defined two categories for time, one tensed and one tenseless. He called the tensed category the A-Series, which included the concepts of *past*, *present* (*now*), and *future*, and he called the tenseless category the B-Series, which included the relational concepts of *earlier than*, *later than*, and *simultaneous with*.

The A-Series is essentially the same as what here we call common time, which might also be called chronological time or clock time. Additionally, the A-series has a *quasi-absolute* element because there was only one occurrence of 10:15AM Pacific Standard Time on April 21, 2025, in San Francisco. There is nothing new or strange about the A-series. However, the B-Series is different, it has no absolute reference point and is independent of when and where.

The B-Series does not include the concepts of past, future, or present and makes no distinction about when or where something happened. Essentially the B-series measures intervals – a beginning and an ending – and determines order and sequence within these measured intervals. For example, a reading of a clock at the start of a scientific experiment is compared to another reading made at the end of the experiment. These *readings* are taken from some sort of clock, which are then inserted into the experimental calculations and interpreted as an *interval of time* – although the actual measurement was a *clock interval* (the difference between two clock readings). The B-series is like the timing of an American football game – where small time intervals accumulate until one hour is reached and the game ends.

---

[68] John Ellis McTaggart, "The Unreality of Time," (1908) *http://www.ditext.com/mctaggart/time.html*.

Of course, today's scientific and technical experimental intervals are most likely measured by a computer's internal clock (the internal cyclical mechanism that all computers use for synchronization).

{**Side Comment on Time and Space Relations**: Regarding space and the location of objects, it must be recognized that there can be no reference to a *there* without a *here*. Likewise, regarding time, there is no *later* without an *earlier*. In the case of the elusive *now*, we sandwich that between the past (memory) and the future (anticipated *now*). That is, both space and time are determined by relational concepts relative to an observer's position and all spatial and temporal statements imply or explicitly state a comparison (there-here, before-after).}

Time Slips Away...

For most of us, time is an unseen and unknown entity, which keeps an indexical pace with events. However, time is actually a convention of convenience that we use in place of change and motion. We can consider money as being somewhat analogous to time. We use money because it is convenient and much simpler than trading in hard goods or services. Today currencies are not backed by precious metals and have value only through common agreement. Currency is simply an indexical measure used to relativize and equate values, and, besides, today most *currency* only exists as computer data.

Having said that, I suppose that I should defend the preceding paragraph. And to start, it seems that our notions of time abstractly symbolize the physical aspects that we associate with change, motion, and concepts such as interval. I say that time abstractly symbolizes these notions because time is not the cause of any of these; it is not the cause of anything, with the possible exception of anxiety and anticipation. Put another way, time is defined through change and motion, not the other way around.

Starting in the late 18th century, and continuing throughout the 19th century, social and economic changes came fast, and as countries industrialized, they rapidly increased their dependence on the clock. As a result, the clock became a dominant factor in how we manage our lives. Under the influence of clocks, people began breaking the day up into smaller chunks.[69] No longer dividing the day into morning, noon, and afternoon, now the common measure of the day is by hour and minute. Instead of meeting someone in the afternoon, it was meeting someone at 3:15PM. All this reliance on clocks changed the way we think about time.

What had been a loose concept based on natural events, became an indispensable concept driven by the needs of modern society. Consequently, our notion of time developed into a dependence on the clock that is both pervasive and deeply ingrained. As previously mentioned, (*Common Time* – page 2) clocks provide a mechanism for dependable repetitive motion and we feel intuitively confidant that these devices keep a steady and invariant pace of measurement.

Is this confidence justified?

When the special theory of relativity (previously mentioned on page 15) demoted time, clocks lost some of their credibility. For example, assume that we have two synchronized clocks setting next to each other, if one of these clocks suddenly accelerates away from the other, the accelerated clock will run more slowly than the one left behind and they become unsynchronized. This is sometimes called time dilation but is actually a slowing down of mechanical movement which is the observed effect of acceleration on mechanical devices that are used to measure intervals that we call *time*.

---

[69] It should be noted that ship navigation was a strong motivator in the development of reliable clocks. The navigators needed these smaller chunks of time keeping - it was the only way navigators could accurately determine the ship's longitude.

For example, consider for a moment the Global Positioning System (GPS) on which so many depend. This remarkable system would not be possible without the knowledge of both the special and general theories of relativity. Indeed, without compensating for these theories, the GPS satellite system would quickly be out of synchronization and become inoperative.

As mentioned previously, irreversible processes, and our common notions about the flow of time present what appears to be a uniform (future-directed) progression of time. Although useful in ordering our lives, this progression, that we call time, is just a mental notion without physical reality. Relativity theory stripped time of its ontological status and demoted time to a mere epiphenomenon. As stated in the above quote by Grandpierre, time is *"a psychological phenomenon with ontological pretensions."*

In everyday discourse we use common time such as when someone asks,*" What time is it?"* or statements such as *"I do not have time for that."* Such references to time are references to a psychological concept that we call time. On the other hand, there are technical references to time that often assume, or give the appearance of assuming, that time is a physical thing. That is, the notion that time has some absolute existence such as the air we breathe. This notion is usually accompanied by the additional notion that time is also a continuum of some kind such that there are no gaps between any two time-like measurements.

Put another way, time is a state of mind that we have learned to believe is something physical and associate it with our *outer intuition* (sense perception) such as the separation of events. The *inner intuition* (mental notion) of time has no geometric image other than that of a line made up of variable length segments that correspond to durations. The physical sensations of events and the ever-changing visual images due to motion define change, which together with our mental notion of a linear sequence convince us that time flows in a definite direction.

It also seems that there has been a long-held belief that change is somehow interconnected or joined with time. That is, many believe that time is the cause of change. Change has many causes, but time is not one of them. This misconception of associating time with change, as well as other misconceptions about time is the result of believing time is a physical entity and part of *concrete* or veridical reality, and even though special relativity successfully refuted the notion of absolute time, the notion of time as something physical appears to be as pervasive as ever – even among scientists who should know better.

Considering time as a requirement or cause of change casts time as some kind of substance with causal capabilities. That is, considering time as possessing some kind of concrescence implies that time is a physical force or causal agent of some kind, even though no physical evidence has ever been found – or will be found.

Change occurs through motion. That is, motion in the broadest sense, which includes everything from the movement of the pen when writing this sentence, to the movement of massive astronomical objects, to chemical reactions such as the rusting of iron, and the radioactive deconstruction of matter at the atomic level.

We know the physical (mechanical) variability of measured time through the theory of relativity, and we know the psychological (mental) variability of the flow of time through personal experiences. The variability of time (or more correctly, the variability of our sense of the passage of nature) is, of course, a well-known experience.

All of us have experienced the fleeting sense of time during pleasant occasions as well as the painfully slow plodding of experienced time during stressful occasions. Of course, the clock on the wall is not bothered by either type of occasion, and its relentless and consistent motion is our main source of an unvarying flow of elapsed *time*. It is through the progression of individual events blending into a sequence of durations, the specious present, and eventually into our sense of the passage of nature that creates our personal notion of time, which is never consistent and steady [Recall the section *Natural Time* - page 4].

*   *   *

Given the many theories and numerous volumes that have been produced dealing with time, it may seem abrupt and presumptuous to dismiss time completely in only a few paragraphs. However, proper reflection on the nature of time makes this the only choice. Philosophers and many others have long debated the reality of time but usually they start with the notion that time is something and then try to prove or justify this choice - the approach herein has been to investigate first then make a choice.

To add to the confusion, even after more than one hundred years of special relativity, many authors of technical papers still have not let go of the notion of time as being something physical. Sometimes the same author will acknowledge that time is a phantom, then later use phrases that imply that time is a physical thing, such as *time itself*, *time beginning or ending*, *static spacetime*, and so on. Other terms used for time include *operational* time, *proper* time, and *physical* time. Additionally, most often, these references to something called time are used without definition such that the authors apparently assume that the reader either knows the definition or can contextually guess the intended meaning.

Understanding the reality of time may not influence our daily activities but when it comes to contemplating fundamental questions about life and the universe the true nature of time must be considered. That is, considering time in the usual (or common) way is a barrier to fundamental thinking about reality and the universe. Changes are essential for advancing our understanding of the natural world and progress only occurs when certain *facts* or *beliefs* are put aside and replaced by novel ideas and concepts.

{**Side Comment on Scientific Time:** Time, as used in science, is a mathematical convenience. For example, time in classical physics is completely reversible. That is, replacement of the variable t [=time] with –t is always permissible, but less often noted is that this t can also be replaced by a function such as t = f(s), where f(s) is an arbitrary function. Here is a quote from a text by one of my former math professors; "*A parametric representation of a curve is not unique. Indeed, are infinitely many representations, as shown by the fact that it may be replaced by an arbitrary single-valued function of a new parameter s, say t = f(s)*..."[70] This is allowable because, when the variable t is used for time in an equation it represents an interval measurement. That is, a B-series measurement of the difference between successive readings of an instrument called a clock. Instead of stating that a certain elapsed time has occurred during the experiment, a function defining the clock's mechanism and mechanical motion could have been used. That is, the measurement of an interval is really a count of the number of clock cycles that occurred. Time could also be replaced by a function representing a linear segment. Since these and other options are cumbersome and less convenient, the variable t is used.}

---

[70] C. E. Springer. Tensor and Vector Analysis with Applications to Differential Geometry, Dover, NY, 2012,13.

Space Reconsidered

"There is no absolute space, and <u>we only conceive of relative motion</u> [emphasis added]; and yet in most cases mechanical facts are enunciated as if there is an absolute space to which they can be referred."[71]

"There is no entity 'physical space;' there is only the abstract space chosen by the physicist as a structure in which to plot phenomena; and some choices give simpler theorems than others (thus making the laws of nature look simpler)." [72]

"Space, like time, would appear to be an abstraction from events. According to my own theory it only differentiates itself from time at a somewhat developed stage of the abstractive process. The more usual way of expressing the relational theory of space <u>would be to consider space as an abstraction from the relations between material objects</u> [emphasis added]." [73]

Maybe these three quotes give away the punchline for this section but, just in case, let us proceed as if they did not and see if these statements are justified.

Over the centuries, space was considered an absolute entity. That is, space was thought to exist as some kind of physical object, a sort of container within which resided the earth and the other heavenly bodies. Of course, if the universe is in a *box*, then what contains this box and so on!

The geometry of this space was the *flat* geometry of Euclid, and it dominated thinking about space for more than two thousand years. This belief in Euclidean space has been so deeply ingrained that it was only within the last two hundred

---

[71]Henri Poincaré, *Science and Hypothesis* (New York: Dover Publications, 1952, English Translation 1905), 51.

[72] Milne 1951, 27.

[73] Whitehead 1920, 37.

years or so that new and radical notions about space emerged that questioned the validity of Euclidean space.

Early in the 19th century, a couple of mathematicians took exception to Euclid's notions of flat space and to his parallel postulate (axiom) in particular.[74] These two mathematicians, Janos Bolyai (Hungarian, 1802-1860) and Nikolai Ivanovich Lobachevsky (Russian, 1792-1856), each, independently, began to develop what we now call non-Euclidean geometry, which is any geometry that deviates from the original Euclidean postulates. What they developed is today called *hyperbolic* geometry and is one of the two most common examples of non-Euclidean geometry. The other common non-Euclidean geometry was developed about the same time by the mathematician Bernhard Riemann (German, 1826-1866) and is called *elliptic* geometry.

Here is a straightforward way to visualize these three types of geometry; think of the Euclidean as the flat surface of a basketball court, elliptic as the surface of a soccer ball (or American football) and hyperbolic as the shape of a Pringle potato chip.

In any case, the invention of non-Euclidean geometry, in the first half of the 19th century, received little attention through the rest of that century. Everyone considered this unusual geometry merely a mathematical curiosity with no connection to the real world. It seems that during the 19th century, when it came to space, scientists were more interested in things like measuring the speed of light and filling space with some mysterious substance called *ether*.[75]

---

[74] This postulate essentially states that given two straight lines, both perpendicular to a third straight line, are parallel in the sense that these two lines will never intersect no matter how long they extend. There are various versions of this postulate but this one works.

[75] This is the so-called *luminiferous aether*; should not be confused with the ether once used as a general anesthetic.

Universal Expansion?

"...if the foundational issues have not been completely clarified from the outset, and if a rigid set of problems has been accepted without enough discussion about their relevance, one's lasting inability to solve some of these problems can be perceived as a symptom of permanent "crisis" ... Yet this so-called "crisis" of the theory may well boil down to a crisis of the too quickly accepted formulation of its problems, or to a crisis of the prejudice associated to it in the name of immediate efficiency [Emphasis added]."[76]

"The observations may be fitted into either of two quite different types of universes. If redshifts are velocity shifts, the model is closed, small, and dense. It is rapidly expanding, ... On the other hand, if redshifts are not primarily due to velocity shifts, the observable region loses much of its significance. The velocity-distance relation is linear; the distribution of nebulae is uniform; there is no evidence of expansion, no trace of curvature, no restriction of the time scale." [Emphasis added][77]

"The assumption that redshifts measure the rate of expansion of the universe is a long extrapolation from the familiar small-scale Doppler effects. Clearly, the data should be analyzed with the aid of as few assumptions as possible. The observations must be corrected for energy effects, regardless of the origin of redshifts. The interpretation of the corrected data is then direct, economical, and very simple." [Emphasis added][78]

The first quote is a warning that premature acceptance and hasty judgments can lead to fundamental issues, which because of these incorrect assumptions and premature acceptances, can lead to permanent creative crisis and

---

[76] Michel Bitbol, "Reflective Metaphysics: Understanding Quantum theory from a Kantian Standpoint," (2010) http://logica.ugent.be/philosophica/fulltexts/83-3.pdf

[77] Edwin Hubble, "Effects of Red Shifts on the Distribution of Nebula," *American Astronomical Society* – Provided by NASA Astrophysics Data Systems. (1936) (http://adsabs.harvard.edu/cgi-bin/basic_connect?qsearch=edwin+hubble&version=1), 552.

[78] Ibid, 553.

underdetermined theories such as those to be discussed in this section.

The second and third quotes are by the American astronomer Edwin Hubble (1889-1953), who was an early 20th century pioneer of extragalactic astronomy. The first Hubble quote was chosen because it mentions the two most common interpretations of the so-called cosmic redshift - velocity and distance - and it shows that, early on (1936), there were serious doubts about the meaning and proper interpretation of the cosmic redshift.

In his second quote, from the same paper, it seems that Hubble's skepticism is even more apparent. He expresses concern about *a long extrapolation, as few assumptions as possible*, and that the data needs *corrections for energy effects*. These seem like the comments of someone expressing both concern and caution when interpreting the redshift phenomenon.

{**Side Comment on Cosmic Redshift:** Redshift is by definition an increase in wavelength and since photon (electromagnetic) energy is proportional to the wavelength, the longer the wavelength means less energy. That is, redshift indicates loss of energy in transit. These shifts in frequency are determined through spectrographic analysis and the resulting spectrographic images rely on the fact that each chemical element (oxygen, nitrogen, etc.) has a unique electromagnetic spectral absorption pattern, and astronomers use these patterns to uniquely identify stellar and galactic chemical composition. If these patterns are out of position, then a shift in photon energy has occurred. Put simply, the light from distant galaxies is compared with light from the normal spectral patterns of light sources here on earth to see if the distant light has been shifted.}

Anyway, before we continue, we need to back up a little. In 1912, the American astronomer Vesto Slipher (1875-1969),[79] while studying spectrographic images from various nebulas

---

[79] In 1912, Vesto Slipher was an astronomer at the Lowell Observatory in Flagstaff, Arizona.

(now called galaxies), was the first to notice the phenomenon that became known as the cosmic redshift. This discovery drew little attention over the next few years, and it was not until better observing instruments became available that the redshift began to receive significant attention. In fact, it was the above-mentioned Hubble, expanding the work of Slipher, who made redshift a hot item.

Fast forward to the 1920s, Hubble used the new 100-inch reflector telescope at Mount Wilson in California to study the redshift phenomenon and by the late 20s had developed what, unfortunately, became known as *Hubble's Law*. This *law*, as originally formulated, related redshift with distance, but, later, to support the claim of an expanding universe the redshift was re-interpreted as a relation between the amount of redshift and velocity.[80] Although, as noted above, it seems that Hubble himself was somewhat uncertain about this idea. In any case, the reinterpreted redshift has played a significant role in the development of what is now called the Big Bang (hereinafter *BB*).

{**Side Comment on Expansion:** The interpretation used by the expansion conjecture is that when the lines are shifted toward the red (lower frequency and lower energy), the object is assumed as moving away, and when shifted toward the blue (higher frequency and higher energy) the object is assumed as moving closer. That is, this conjecture is analogous to the well-known doppler effect that we experience with sound waves.}

The expansion and *BB* conjectures started in 1927, when Georges Lemaître (Belgian, Catholic Priest and physicist, 1894-1966), used the redshift as a velocity measure to suggest that the universe was expanding. Later, in 1931, Lemaître developed this expansion conjecture into the idea that the universe started with an explosion of something he called a

---

[80] Along with other problems, this reassociation committed the logical fallacy of *concurrence does not imply causation* or invalid *cause and effect* error.

primitive atom. This was the beginning of universal expansion and the *BB conjecture*, although it was not until 1949 that the term *BB* was first used. English astronomer Fred Hoyle coined the term *Big Bang* during a 1949 BBC radio broadcast - it was meant to be derogatory.

Over the years, the *BB* and expanding universe conjectures (hereinafter *BB/E*) gained in popularity to the point of being considered by many, if not most, astronomers and astrophysicists as a confirmed reality. On the other hand, there is still a large and possibly growing group of skeptics who are not sure all is right with this concept. This is because, over the years, the *BB/E* conjecture has kept running into trouble. However, instead of looking in a new direction, it has always been decided that all the *BB/E* needed was another patch.

For example, in the 1970s, there was a realization that the current *BB/E* narrative did not *grow* the universe in quite the right manner. Specifically, it was realized that very soon after the BB/E some fine-tuning was required for the universe to appear the way it actually does. However, since some possible choices for fine-tuning might have certain unwanted philosophical implications, the *right kind* of tuning had to be *discovered*.

Then along came Alan Guth[81] who, in 1980, suggested that almost immediately after the *BB/E*, the universe began to expand much *faster than the speed of light,* that is remarkable enough but even more remarkably, expansion lasted only about *one ten thousandth of a second*, then the *cosmic inflation* (as it came to be known) suddenly stopped as it reached just the right degree of expansion!

How convenient, a perfect fit! This convenient *discovery* (made-up story?) saved the day for the *BB/E* enthusiasts and Guth's idea was quickly accepted as gospel. This was so even

---

[81] In 1980, Alan Guth was an astrophysicist at MIT, whose prior notoriety was his being voted as having the messiest office in Cambridge.

though Guth's proposal strongly violated the special theory of relativity provision that the speed of light could not be exceeded by any particle that has rest mass greater than zero (let alone the whole universe at that time!)![82] So, the BB/E notion was patched up again and with the addition of cosmic inflation it grew in scope to become the BB/EI notion.

However, the truth is, that the introduction of inflation was hand fixing and fine-tuning to the extreme, but somehow it managed to become widely accepted as an integral part of modern cosmology. Moreover, the inflation notion also included a generating elementary particle called an *inflaton*, which, not surprisingly, is yet to be discovered. It is easy to picture this ongoing search for the mythical *inflaton* particle with a bunch of latter-day *Argonauts* scurrying about in all directions searching for the *golden fleece* of inflation.

Perhaps the biggest problem with inflation is the dependence on the BB conjecture, which is itself an unproven notion and, as of today, there are *no confirming observations of expansion,* and the privileged BB/EI is allowed *ad hoc* corrections that other theories are denied. These are only a few examples of the many problems encountered and the *by-hand* fixes that, over the years, have been applied to *BB/EI*.

As another example of questionable conjectures that are claimed to support the BB/EI, there is the much-touted cosmic microwave background radiation that is claimed to be *proof* of the big bang even though it is possibly just *intra-galactic* and not *extra-galactic* as claimed.

The external world is presented to us much like a shadowy reflection of reality, which presents only a specific perspective. This is not to say that we live in the shadowy world of a *Platonic cave,* but that our limited ability to perceive reality

---

[82] In special relativity theory, the rest mass is essentially equivalent to what we normally call mass for an object at rest (not moving relative to its surroundings – such as a coffee cup sitting on a table).

constrains our knowledge of the actual world. Just as we only see the surface of a body of water, we know that there is a vast reality hidden below. Unable to see below, we make assertions about this hidden reality based on interpretive extrapolations that go far beyond the given. Together the BB/EI, and other conjectures - such as dark matter and dark energy[83] make up what is called the *Standard Model of Cosmology* (hereinafter SMoC), which appears much like a leaky boat that has been patched too many times.

For example, physicist Pavel Kroupa (Czech Australian astrophysicist, 1963 -) wrote a paper a few years ago that clearly illustrated the problems with the SMoC.[84] In this paper, he outlined 22 issues with the cosmology model that had arisen over the 32 years between 1980 and 2012. In this paper, Kroupa clearly identifies these many issues with the SMoC, which are either ignored or, as is sometimes the case, altered through spinning to look like a discovery. Examples of the latter include dark matter, dark energy, and inflation, which were all put in by hand to solve problems, but their *discoveries* (patches) were touted as worthy scientific achievements. Here is a quote from his summary: *"Put another way: that the currently popular SMoC needs to be based to more than 96 per cent on unknown physics is nothing more than an expression of our present-day ignorance of how mass, space and time unify, i.e., of cosmological physics."*[85]

Even with all these fixes, other factors seem not to have been completely considered, including; density of photons, loss of photon energy (redshift = tired light?), and the

---

[83] Dark matter and dark energy are hypothetical stuff that some say may represent as much as 85% of the total matter of the universe. Both were added into the SMoC – by hand - to patch it up. Neither have been found after many years of investigation.

[84] Pavel Kroupa (2012), "The Dark Matter Crisis: Falsification of the Current Standard Model of Cosmology." https://arxiv.org/abs/1204.2546v2

[85] Ibid, 36.

unknown impact of a multi-billion year journey on light propagation.[86]

The SMoC is composed of too many assumptions and too many unverifiable conjectures such that skepticism about the veracity of modern cosmology seems rationally unavoidable. Modern cosmologists might look with amusement at the epicycles and geo-centric cosmological conjectures of the past, but are today's conjectures any better? It appears that much of current cosmology is more like fanciful or wishful thinking than science.

The historical track record for astronomy and cosmology is not very convincing. It is exceedingly naïve and arrogant to claim that in a very short time; we have – finally - been able to *get it right* and that the current SMoC is a *true and accurate description* of the origin and development of the universe.

We need only consider the history of cosmology to realize that, although there is a lot of data being collected and a lot of analysis, and even though cosmologists have been wrong over and over again, all this observation and analysis is followed by new notions that are then claimed as true and accurate. As someone once said; *"Cosmologists and astronomers are often in error, but never in doubt."* True believers in SMoC seem to live in a fantasy world where they apparently really believe their tales of exactly what happened and when it happened, but without a clue as to *why* it happened.

Beyond these doubts about the veracity of current cosmological conjectures are doubts about whether we are even capable of making sound cosmological conjectures. This doubt has been around for more than a hundred years, and these doubts were voiced by some of the abler thinkers of their times.

---

[86] Those interested in additional comments should refer to: Hartnett, J. G. (2011): "Is the Universe Really Expanding?" https://arxiv.org/abs/1107.2485

For example, the famous Austrian physicist and philosopher Ernst Mach (1838-1916) held the view that the universe was not an object to which we could apply meaningful statements. Another example is that of the Pierre Duhem (French physicist and philosopher, 1861-1916) who believed that physical theories about the universe could never be confirmed or negated.[87] Examples that are more recent include British Astronomer M. J. Disney,[88] South African cosmologist George Ellis,[89] and the above-mentioned Pavel Kroupa. This list could go on, but this is enough to make the point that even today not all scientists are *true believers*.

As discussed above, astrophysicists have been putting patches on the *BB/EI* conjecture for many years and they seem determined to save this conjecture at any cost. Whether this is because they believe the conjectural basis is fundamentally sound or there are other reasons is difficult to say. However, possibly (probably?) these extreme efforts to save the *BB/EI* conjectures are due to the influence of socially motivated hidden agendas.

There seems little doubt that the *BB/EI* and associated conjectures are, at least to some degree, ideologically motivated and this motivation is inclined toward reductionism and scientific realism. Despite all these attempts to rescue the conjectures using *ad hoc* patching, the BB/EI conjecture still fails to have predictability or any predictions that can be tested.

---

[87] The Mach and Duhem examples are taken from Helge Kragh, "The Controversial Universe: A Historical Perspective on the Scientific Status of Cosmology", https://d-nb.info/101824364X/34.

[88] M. J. Disney (2000), "The Case Against Cosmology", arXiv: astro-ph/0009020v1.

[89] George F. R. Ellis (2006). "Issues in the Philosophy of Cosmology. Section 2.8.1, Misconception 1," arXiv: astro-ph/0602280v2.

It should be noted that in discussions about the expansion of the universe, it is often stated that expansion is due to the expansion of space itself, not due to relative motion of spatial objects. This obviously assumes that space is something that can be expanded. That is, a physical entity of some kind. This is also sometimes given as the reason there is no center of expansion.

To illustrate this last point, a two-dimensional model is sometimes used. This is the somewhat famous balloon model, where you put a few dots on a partially inflated balloon then continue to inflate the balloon such that all the dots will move away from each other without having a center of expansion. Very tidy but, as we will see later, this model relies on hidden structure.

In a recent publication physicist Lee Smolin (American, 1955-) stated that the BB/EI is "an established scientific fact, which has been elaborated into a detailed story..." and that "we have detailed theories [BB/EI] that pass numerous observational tests."[90] To say that BB/EI fit observations is correct, but it is correct because the BB/EI was designed to fit observations, which is ad hoc fixing to the extreme and is entirely inappropriate since its only justification is to save their conjectures! This reasoning is typical of the reductionist and scientific realism agendas that are now trying to redefine the notion of what is a valid scientific theory to include such wild fantasies such as the multiverse [discussed in the next section].

In any case, since the 1930s, hundreds of books and thousands of technical papers have been written about the redshift, expanding universe, and the big bang. Therefore, to

---

[90] Lee Smolin, "The Big Bang was the First Moment of Time," *This Idea Must Die*: Edited by John Brockman, (Harper Perennial, New York, 2015), 32.

keep this discussion from *inflating* out of control and getting too deep into the weeds I will close with a few final comments.

To be sure, verification of any cosmological conjecture is difficult, but if they rate consideration as true scientific theories, then verification beyond mere wishful thinking and hand fixing is necessary. Unfortunately, modern cosmology seems more interested in saving the SMoC than in truth. To proclaim this pieced together patchwork, as being proven reality is a stretch, at best, and an outright counterfactual fabrication, at worst.

Cosmology, which has somehow gained more credibility than deserved, is not a physical or quantitative science in the sense that chemistry, biology, physics, and so on are because cosmology is strictly an observational practice with few experimental possibilities. Practitioners build toy models - such as the well-known Friedmann–Lemaître–Robertson–Walker (FLRW)[91] model - that are used to claim that we know when the universe started, how it developed, and a lot of other stuff based on interpretations of observations that always lend themselves to more than one valid possibility. There are huge differences between experimental science and observational science that put sever limits on cosmology as a true scientific discipline.

Adherence to the BB is clearly an example of groupthink such that these astrophysicists and cosmologists are determined to hang on to this notion and the notions of inflation and accelerating universal expansion at any cost and no matter how many hand-fixes are required. Put simply, it is likely

---

[91] The FLRW is a toy model metric, developed in the 1920s and 1930s by the named authors, that many believe represents the structure of the universe. The metric depends on the key assumptions of universal homogeneity and isotropy in the distribution of matter (galaxies, clusters of galaxies, etc.). It is part of the SMoC.

they will continue to play around with their toy models of oversimplification and *ad hoc* patches.

Choosing a particular cosmological model is more psychological and sociological, and perhaps even political, rather than scientific. This is partly because all models require huge extrapolations and assumptions based on observations that leave considerable room for multiple interpretations. And some of these extrapolations and assumptions (such as *inflation*) have no direct observational support – or indirect support either.

Today's science seems to be making the same mistakes made in the 18th and 19th centuries with the development of classical physics. Like those developers of classical physics, today's theorists seem to believe too soon that they have achieved their goals and cling to these theories with a religious tenacity that proclaims to have arrived at universal truths about the universe.

To end this section here is a quote that lucidly states the case: "More specifically, our goal is to show that (1) excluding certain pathological examples, every cosmological model of our universe is empirically underdetermined;[92] no amount of observational data we could ever (even in principle) accumulate, can force one and only one cosmological model upon us. Additionally, we also claim that (2) even if one assumes a principle of uniformity – that the physical laws we determine locally are applicable throughout the universe – these general epistemological difficulties remain."[93]

---

[92] For those who might be unfamiliar with the word underdetermination here is the wiki-def: "In the philosophy of science, underdetermination refers to situations where the evidence available is insufficient to identify which belief one should hold about that evidence."

[93] John Byron Manchak. "Can we know the global structure of spacetime?": http://philsci-archive.pitt.edu/4162/.

The Multiverse and the Fine-Tuning Argument

'The flights of physicists and astronomers today answer <u>the es-thetic need for satisfaction of the imagination rather than to any strict demand of unemotional evidence of rational interpretation</u> [Emphasis added].'[94]

When I started writing about the BB/EI, I decided not to mention the so-called multiverse notion (or the many-worlds - or Everett – interpretation, which definitely will not be discussed). However, since there seems to be growing interest in this notion about an uncountable number of universes, a few comments seemed appropriate.

The multiverse notion essentially states that our universe is only one of a very large or even an innumerable number of universes, and that the one we live in was *randomly conditioned just right* for life. This notion – generally referred to as the *Fine-Tuning Argument* (hereinafter FTA) claims there is no fine-tuning (intentional or otherwise) of various constants of nature in support of life, but that it was only random chance that this universe just happened to have the correct parametric values (physical conditions) for life. Lucky us?

Advocates of the FTA, like Andrei Linde (Russian American Physicist, 1948-) claim that even though the multiverse is neither refutable nor irrefutable, it is based on the Standard Model of Cosmology (SMoC – page 71), so therefore, it must be correct. He also stated that any argument against the multiverse must show that only one universe is possible.[95] *This is obviously backward thinking; the burden of proof is necessarily on the advocates of the multiverse, not the other way around!*

* * *

---

[94] Dewey, (1934) 2005, 31.

[95] Andre Linde, "The Uniformity and Uniqueness of the Universe," *This Idea Must Die*: Ed. John Brockman, (Harper Perennial, New York, 2015), 44.

There must be some demarcation between science and non-science, and to say, as some scientists and philosophers do, that testability is naïve and amateur philosophy is not acceptable as an excuse to dismiss the notion that a valid scientific theory must have true testability and make predictions not yet observed. That is, any notions that do not have these features are attempts at diluting scientific evidence such that it is no longer science but a wishful fantasy. Another possibility, although weaker, is that given the correct circumstances, the ability to explain previously observed, but unexplained, anomalies might constitute a valid scientific theory, if, and only if, at the same time, novel and/or testable predictions are possible.

Recently it came to my attention that in 1986 there was a Supreme Court case concerning creationism versus Darwinian style evolution. This case was settled in large part based on the written testimony *presented under the signatures of 72 Nobel Prize laureates* in support of evolution and extremely critical of creationism. Here is a quote from that written testimony:

"To be a legitimate scientific "hypothesis" an explanatory principle must be consistent with prior and present observations and must remain subject to continued testing against future observations. An explanatory principle that by its nature cannot be tested is outside the realm of science. The process of continuous testing leads scientists to accord a special dignity to those hypotheses that accumulate substantial observational or experimental support. Such hypotheses become known as scientific 'theories.' If a theory successfully explains a large and diverse body of facts, it is an especially 'robust' theory. If it consistently predicts new phenomena that are subsequently observed, it is an especially 'reliable' theory. Even the most robust and reliable theory, however, is tentative. A scientific theory is forever subject to reexamination and -- as in the case of Ptolemaic astronomy -- may ultimately be rejected after centuries of viability. [emphasis added]"[96]

This statement may have won the case for anti-creationists, but it appears to spell disaster for the multiverse (and the SMoC), since the multiverse notion does not meet any of the criteria ascribed by these 72 Nobel laureates, and others, since it is completely devoid of possible observation and impossible to test experimentally! If that is not convincing enough, here is what the famous physicist Richard Feynman had to say about what constitutes science: *"The principle of science, the definition, almost, is the following: The test of all knowledge is experiment. Experiment is the sole judge of scientific 'truth' [emphasis in the original]."*[97] Again, here is a strong statement of the indispensable requirement of testability if a notion is to be considered scientific. And here is one more quote by Henri Poincare:" *Experiment is the sole source of truth. It alone can teach us something new; it alone can give us certainty. These are two points that cannot be questioned."*[98]

As another example, according to Bernard Carr (British astronomer, 1948 -), just as historical beliefs about the design of the universe went from the claim of the universe being geocentric to the later claim of being heliocentricity, the multiverse is just the next step in the process.[99] However, there is, at least, one major problem with his statement, it ignores the fact that the other changes mentioned were due

---

[96] Edwards v. Aguillard: U.S. Supreme Court Decision. Brief submitted under the signatures of 72 Nobel Laureates, 17 State Academies of Science, and 7 other scientific organizations in support of the appellees. www.talkorigins.org/faqs/edwards-v-aguillard/amicus1.html

[97] Richard Feynman, *The Feynman Lectures on Physics: Volume* I. (Addison-Wesley, Reading MA, 1989), 1-1.

[98] Poincaré (1952), 140.

[99] Bernard Carr, as quoted in Tim Folger, 'Science's Alternative to an Intelligent Creator: The Multiverse Theory', *Discover Magazine* (December 2008), online version.

to improved observational instrumentation and analytical analysis, while the multiverse has no observational support regardless of instrumentation, no possibility of analytical analysis, or any basis that can honestly be considered a fact. Obviously, the multiverse conjecture is a notion that can never be proven or dis-proven through observation or any other means.

True advances in science have not been made by modeling data, but by intuitive thinking followed by conjecture with subsequent attempts to verify the predictive and explanatory capabilities of the conjecture through experimental testing. The multiverse meets none of these criteria and has no predictive capability.

Linde and the other advocates of this fantasy are correct when he says that the multiverse notion is irrefutable, so is the existence of the *tooth fairy*. It is also non-measurable in the sense of observational confirmation or any empirical verification at all. The multiverse and BB/EI together amount to a *theory of anything and to predict anything is to predict nothing*!

Put another way, the multiverse/FTA argument essentially states that the physical parameters that govern the physics of the universe were randomly (accidentally?) fine-tuned for life, and as discussed above, this is claimed to imply the existence of an innumerable number of other universes, each with different parameters and parametric values that are *somehow* randomly selected. These parameters are the 26 or so constants of nature that are part of the Standard Model of Particle Physics.

Arguments for fine-tuning claim that the existence of a universe that supports life is a probabilistic issue and that as stated by philosophy professor Peter Epstein (UC Berkeley) *"...conditions needed to support life are extremely unlikely to occur in any given universe according to our best cosmological theories*

[Emphasis added]."[100] Since it is impossible to know anything at all about *any given universe*, we can only know about this universe such that making statements about any other universe is essentially fanciful wishful thinking and is certainly not science or even good philosophy!

On the same page, Epstein goes on to say; "...*the chances of there being some life supporting universes are higher the more universes there are...*" How can someone make such statements when 1) there is absolutely no substantive evidence that even remotely suggest the existence of other universes, 2) This statement pre-supposes a knowledge of what constitute a life viable universe about which we know almost nothing beyond speculative imagination, 3) his argument about the more universes the more likely to have at least one supporting life is a perfect example of the *Gambler's Fallacy* [discussed below]. The rest of Epstein's paper is essentially a continuation of unsubstantiated and unverifiable statements about probabilities, likelihoods, and so on. It seems that Epstein, like many others, is willing to go to any extreme of imagination to avoid anything that might even hint at purpose or intentionality within the universe.

Furthermore, with regards to these physical parameters that are claimed to determine the structure and, indeed, the very existence of the universe, I believe that it can safely be said that no one knows enough about the interaction of these parameters to really make claims about the possibility of being compatible with life. In support of that statement, here is a quote from a recent (February 2019) paper that discusses both the multiverse and the fine-tuning argument:

"In particular, the range of possible parameter values (the minima and maxima *of the distributions) must be specified. A full assessment of fine-tuning requires knowledge of these fundamental probability distributions, one for each parameter of interest (although they are*

---

[100] P. F. Epstein. "The Fine -Tuning Argument and the Requirement of Total Evidence", *Philosophy of Science Journal Volume 84*, 2017, 640.

not necessarily independent). <u>Unfortunately, these probability distri-</u> <u>*butions are not available at the present time*</u> [emphasis added]."[101]

This statement in itself is a rare acknowledgement of theory underdetermination. Put simply, this means that permissible parameter values are not known, and how these parameters may vary and interact is not known, so any attempts to model possible universes by manipulating these parameters is simply guess work.

Wait, there is more, this review actually admits that the multiverse notion is simply a fantasy:

"The consideration of possible alternate universes, here with different incarnations of the laws of physics<u>, is by definition a counterfactual enterprise</u>. This review considers the ranges of physical parameters that allow such a universe to be viable. <u>Since alternate universes are not observable, this endeavor necessarily lies near the boundary of science</u>. Nonetheless, this discussion is useful on several fronts: First, one can take the existence of the multiverse seriously, <u>so that other universes are considered to actually exist</u>, and the question of their possible habitability is relevant [emphasis added]."[102]

Here is a candid admission that the multiverse notion is pure fantasy. Again, put simply, the notions put forward concerning the multiverse are simply fictitious and contrary to the second emphasized statement that the multiverse notion, is not just *near the boundary of science*, but does indeed lie outside the boundary of science, and inside the boundary of fantasy, and the last statements underlined in the quote is really amazing since it states that even though the multiverse notion is make-believe, *we can pretend that it is fact*! This paper continues this fantasy narrative for more than <u>200</u> pages pretending that something significant could actually come out of this fictional story.

---

[101] Fred C. Adams. "The Degree of Fine-Tuning in our Universe – and Others", https://arxiv.org/abs/1902.03928 (Feb 2019), 12.

[102] Ibid, 12.

Some attempts at explaining fine-tuning sometimes try to analogize with some type of random process, such as a giant billion slot roulette wheel, catching the largest fish in a lake, or some other mechanism of apparent random selection that gets it just right. These arguments are simply weak attempts to explain in reductionist terms how our universe just happened to support life. Also, they never seem to explain who or what spins the roulette wheel or catches the fish.

On the other hand, it is surprising that sometimes we see references that imply purposeful tuning from those who usually denounce such notions. Here are two examples from well-known scientists – Lee Smolin and Stephen Hawking:

"The existence of stars rests on several delicate balances between the different forces in nature. These require that the parameters that govern how strongly these forces act be tuned just so. "103[Emphasis added]

"The remarkable fact is that the values of these [fundamental] numbers seem to have been <u>very finely adjusted</u> to make possible the development of life."104[Emphasis added]

These two phrases *tuned just so* and *very finely adjusted* certainly seem to imply a non-random process, which implies purposeful and intentional action – however, no suggestions seem to have been made about who or what did the fine-tuning.

As mentioned above, some of these arguments are victims of the *Gambler's Fallacy* and others the *Inverse Gambler's Fallacy*. The Gamblers Fallacy, for example, is the assumption that after a long sequential series of *heads* appearing with the flip of a coin, that the odds improve for getting *tails*

---

103 Lee Smolin. *The Life of the Cosmos*. N Y, NY: Oxford UP, 1997. 37.

104 Stephen Hawking. *A Brief History of Time: From the Big Bang to Black Holes*. New York, NY: Bantam Books, 1988, 125.

on the next toss, when, in fact, the probability does not change for the next flip. On the other hand, the Inverse Gamblers Fallacy is the mistake of inferring a large number of trials (such as a lifetime's worth of poker hands) solely from observation of a single success (such as a royal flush). As another example, winning a lottery does not mean that a person has played the lottery many times, it could just as well be their first try.

Arguments about fine-tuning seem, for the most part, to miss the point. Life exists so quibbling about whether it is rare, unique, or abundant is essentially a philosophical question, not a scientific question. It is also questionable whether any discussion about fine-tuning should only refer to micro-features defined in the Standard Model of Particle Physics.

Instead of asking why the *laws* of nature seem so *fine-tuned* for the support of life, perhaps a better question is how did life manage to conform to the existing laws of nature? That is, it seems that life is fine-tuned for compatibility with the laws of nature, not the other way around. Since the environment existed before life, it is trivially obvious that life found a way to exist within this environment not the other way around. By extension, it seems reasonable to assume that since life can and did occur here on earth, then life can and did occur on a myriad of other planets as well.

It seems strange and hard to understand how reputable scientists can state, seriously, that the FTA or any other argument confirms the multiverse notion. Perhaps, as I personally believe, it is because of hidden agendas that these scientists support the FTA and the multiverse notions as a means of suppressing any non-materialistic beliefs.

As discussed, there are sound rational, logical and physical reasons to reject both the notion of the multiverse and the FTA, while, conversely, there are no sound rationally logical reasons to accept them. Given this, it is clear that the multiverse and the FTA should be relegated to the world of science fiction and fantasy.

All this stuff reminds me of the word *verisimilitude*, which is defined as - having the appearance of being true such as a fantasy novel. It seems apparent that either knowingly or unknowingly this technique of 'what if's and pretend' is used by those with materialistic inclinations and agendas to provide a picture that appears true but is actually made up and only superficially factual such that on close examination is seen for what it really is.

As a final comment on this it should be noted that some of these *reputable* scientists are using US taxpayer money from the National Science Foundation and other tax-based sources to promulgate these fictitious notions under the guise of being scientific!

Universal Motion

"The experiences which prove the dynamics in-equivalence of different states of motion teach us that the world bears a structure. However, in the concept of absolute space this inertial structure is evidently not sized up correctly; the dividing line does not lie between rest and motion but between uniform translation and accelerated motion. [Emphasis added]"[105]

As this quote implies, all matter is in constant motion and there is only the question of whether this motion is uniform translational motion or accelerated motion.

Although the terms uniform translation or accelerated motion may seem unfamiliar, we have all experienced motion of both types. Uniform translational motion is experienced when riding in a vehicle that is moving at a constant velocity. More technically, an object moving with uniform translational motion is said to have a *comoving frame of reference*. That is, using the example of an automobile moving at constant speed,

---

[105] Hermann Weyl, *Philosophy of Mathematics and Natural Science* (Princeton, NJ: Princeton University Press, 1949), 101.

the motion of the vehicle is only known by observing the apparent motion of objects outside of the vehicle. Put another way, the passengers and vehicle share a common frame of reference (set of coordinates), and this is why they do not feel any sense of movement and are aware of their movement only by observing the flow of objects by looking out a window.

On the other hand, it is only through accelerated motion that there is anything like a *non-comoving frame of reference* and that is the system referencing the body being accelerated. Put another way, accelerated motion has an absolute aspect as *felt* by the object being accelerated.

Although a change in velocity may be observed by an observer outside of the moving object, the absolute aspect of acceleration is only applicable to the body being accelerated, not to any observer of that acceleration, who only see an increase in separation and rate of separation. That is, accelerated motion may be observed through change in velocity with regard to the observer, but only the object being accelerated *feels* the acceleration, which is due to inertial mass resisting the accelerating force.

Accelerated motion comes in two types. The first is the linear acceleration that we experience when a motor vehicle accelerates in a straight line. Of course, the same can be said about decreasing relative speeds where an object experiences deceleration. Both types are familiar to anyone who has traveled in a commercial airplane where the acceleration at takeoff and the deceleration on landing are very apparent.

The second type is angular acceleration, which is experienced when a vehicle is turning a corner.[106] That is, the mass of an object in uniform translational motion feels no resistance to its motion, while, on the other hand, an

---

[106] This angular acceleration is independent of whether or not the vehicle is moving at a constant speed. Angular acceleration is also experienced by an object rotating about an internal axis such as a spinning top or the daily rotation of the earth.

accelerated object *feels* a reaction proportional to the object's mass and the force causing the accelerated motion.[107] As mentioned, uniform motion is only detectable through observation of a *non-comoving frames of reference*, such as the view of objects outside a moving vehicle. Accelerated motion is self-realizing within the objects coordinate system.

For humans, the feeling of acceleration is partly due to anatomical organs that are sensitive to changes in our motion relative to the earth's surface. Put another way, this sensation of acceleration that we commonly feel is due to the inertial effect on matter when the rate of motion changes relative to what was, until the change in velocity, a co-moving or inertial frame of reference – the earth.

This brings us to the topic of this section, which is universal motion, or, simply, the notion that all matter (as defined on page 7) is always in relative motion – either uniform translational or accelerated (linear or angular). If this notion of constant or universal motion seems an overstatement, then consider the reverse and ask what matter is not in motion and when is this matter motionless.

To be sure, whether it is the constant orbital and rotational motion of our planet all earth-bound matter is always in relative motion. Obvious examples include the apparent daily motion of the sun and the nightly apparent rotation of the stars (except one). More subtle examples include the motion within a glass of water that has been left undisturbed for some time. The water seems perfectly still, but we know that

---

[107] Force is equal to mass multiplied by acceleration ($F = ma$). Put another way, acceleration (**a**)equals the applied force (**F** - such as gravity) divided by the mass (m) of the object being accelerated, or $a = F/m$.

the water molecules are constantly bouncing around in a motion called Brownian movement.[108]

Although, at the level of everyday experience, other agents play a role, the dominant agent, for the universe as a whole, is gravity. Even though gravity is many orders of magnitude weaker than other forces such as electromagnetism, gravity has the overwhelming advantage that it is pervasive and cannot be blocked or cancelled.[109]

When considering concepts such as mass, inertia, and motion, it should be kept in mind that there are no physical objects that are entirely force-free. Only idealized mathematical abstraction or other mental constructions can be said to be force-free. Objects in a state of equilibrium regarding their local frame of reference are in this state by balanced forces acting on that object. For example, the rock setting on the ground pushes against the earth with a force that is equal to the force that the earth is pushing against the rock.

One can imagine gravity as that persistent and irresistible force that is always active and from which there is no escape. The weightlessness or *free-fall*, as experienced by orbiting astronauts, is not free. It is a delicate balance and constant struggle between gravity and the angular acceleration of the orbiting object. Gravity will not go away!

{**Side Comment on Gravity**: Another interesting fact about gravity is that all earthbound objects accelerate at the same rate if air resistance is eliminated such as in a vacuum, where a rock and a feather will fall at the same rate. It is indeed quite remarkable that, in a vacuum all

---

[108] This phenomenon was discovered by Robert Brown (1773-1858) in 1827 while experimenting with some pollen suspended in water. It clearly demonstrated that the molecules of water were in constant motion even when left undisturbed.

[109] The gravitational constant is 10^{-36} (= 0.000000000000000000000000000000000001) times weaker than the electromagnetic constant!

objects fall at the same rate– that is, any physical body in the same gravitational field will have the same acceleration such that it is not the objects mass that determines the rate of acceleration but only the fact it is a material object of some kind. We know a lot about gravity, but we do not know how or why this equal acceleration works.}

{**Side Comment on Universal Change:** An adjunct to, and concomitant with, universal motion is universal change. There is a continuous and irresistible cycle from undifferentiated substance into differentiated substance and back again into undifferentiated substance. Universal change is an enfolding and unfolding, nothing remains unaltered. For example, undifferentiated non-living substances are enfolded as differentiated living organisms, then eventually unfolded back into undifferentiated substances.}

Discrete versus Continuous Motion

Given universal motion, there remains the question of whether this motion is discrete or continuous? Like other questions examined herein, this is an ancient question, whose origin is lost in antiquity. However, we do know that it dates at least as far back as ancient Greece and notably to the paradox-like scenarios attributed to Zeno of Elea (490–430 BC).

Zeno's paradoxes rely on the notion that an object would have to pass through an infinite number of positions when moving from some position A to another position B. Zeno speculated that movement through an infinite number of intermediate positions would require an infinite amount of time and, therefore, motion does not exist and is an illusion.

One of the problems with this paradox of infinite positions requiring infinite time is the category mistake of applying *physical reality* to an *abstract mathematical concept*. A category mistake is something like comparing apples and oranges, or, since apples and oranges are both fruit, more like comparing apples and rocks. In this case, the category mistake is an improper comparison of an assumed physical continuum (infinite number of physical positions) and an abstract continuum (mathematical infinity).[110] Or put another way, equating a mathematical abstraction (infinite number of points on an idealized line segment) with a physical length or distance between two objects is an invalid comparison.

It is sometimes said that Zeno's paradox is solved through a finite sum of an infinite number of terms. This makes a good point, however; this requires that the infinite series of terms converge to a finite number, which is not always the case.[111]

Fast forward about 2400 years and it turns out that all this is connected to the so-called *quantum of action* or Planck's constant (usually designated by the letter h). This was the discovery of the physicist Max Planck (German, 1858-1947) who, in 1900, published a paper demonstrating that at the microscopic level energy was incremental. This level, called the Planck scale, is too small for direct probing and is so removed from phenomenal experience that it has been said that physics meets philosophy at the Planck scale.

According to Planck's theory, the smallest possible increments of energy are proportional to his quantum of action, and, by implication, the Planck scale represents the smallest possible increments allowed by nature. This scale includes values, representing time (in the sense of event separation), length, and mass (energy), which respectively represent the smallest possible increments – for each type of measure. This means that at the Planck scale motion is incremental. This, in turn, means that between any two physical positions (such

---

[110] Within mathematics, there is something called the *continuum of numbers*, which, put simply, is the question of how many points are on a Euclidean line of any finite length.

[111] It should be noted that the sum of an infinite series is never calculated using an actual infinity of terms but approximates an infinite number of terms through algebraic manipulations.

as this line segment ----------) there is always a finite number of increments.

OK, maybe this seems to be stretching it a bit. The motion we commonly experience certainly seems continuous and smooth, but that is just because of our limited perceptual capabilities. Obviously, we cannot view motion at the Planck scale, just as we cannot see the individual molecules of flowing water or the discrete molecules of any surface that seems both smooth and continuous. For example, the paper that I am now writing on feels smoothly connected but if I were to look at it through a microscope it would appear extremely lumpy and rough.

So, what is the point? The point is that, as with other notions discussed herein, a deeper understanding of our natural world increases our appreciation of the wonders of nature and enhances our worldview. Although not commonly important in everyday situations, discrete versus continuous motion is important when scientifically or philosophically contemplating the fundamentals of nature.

When Planck developed the notion of quantum, it was a leap of imagination that shook the scientific world to the most fundamental level. More than just a paradigm shift, this notion that nature, at its most fundamental level, was discontinuous and not continuous was truly revolutionary and counterintuitive to an extreme. It was difficult enough to consider the notions that energy at the smallest level was discontinuous, but that matter, motion, and light were also discontinuous was even more difficult to accept.

This notion came at a time when the atomic nature of matter was still being debated and definitely made the atomists position much stronger, but the *continuous faction* was still not finished off.[112] That finish came a few years later by two

---

[112] Four years earlier when radioactivity was discovered the position of the atomists was all but confirmed.

papers that Einstein published in 1905, the one on Brownian movement and the one on the photo-electric effect. The first, convinced most skeptics that matter was atomic in nature, while the second reinforced Planck notion of energy packets and, although not realized at that time, was the beginning of learning about the interactions between photons and electrons that eventually grew into the theory of quantum electrodynamics.

Because of Planck's discovery, the year 1900 might well be considered a point of demarcation between the classical scientific world and the new physics that turned classical science on its head. Unexpected discoveries such as the constancy of the speed of light, the strangeness of black body radiation, and the new discovery of the abnormal orbit of the planet Mercury were mysteries that needed solving. It turned out that these three anomalies were resolved by Max Planck, Albert Einstein, and others over the next twenty or so years. However, science was to pay a price for these discoveries in the form of new and strange paradigm changes that even today continue to confound intuitive thinking about the natural world.

{Side Comment on Continuity versus Continuous: Nature has continuity, but nature is not continuous. Discrete events, such as the frames of a video that are systematically ordered have continuity, but they are not continuous in the sense of having no gaps or interruptions. Continuity does not, necessarily, mean seamless connectedness (continuous), but is a reference to a certain sense of order.}

{Side Comment on the Planck Scale: Currently, particle accelerators probe into the micro-world at the level of about 10^{-17} centimeters, which is about halfway to the ultimate smallness of the Planck length of about 10^{-33} centimeters. As pointed out by D. Bohm (American physicist, 1917-1992), we should consider that since the orders of magnitude of complexity between the macro-level of human experience and the smallest level yet probed by science is only about halfway to the

Planck level, this leaves plenty of room for a great deal of undiscovered complexity and hidden reality.[113]}

The Infinite

Infinite (or infinity) is a word that has been much used and much abused to the extent that most usage of this word is either imprecise or just plain wrong. Moreover, it is safe to say that no one who uses this word has ever experienced infinity in any form. In fact, the infinite, if it exists in any form experienced by humans, exists only as an abstract mathematical convention that was mentally conceived as a calculational convenience.

Infinity is a label we attach to the notion of something increasing or decreasing without limit. That is, the term infinite is not only used for something increasing without limit (an *exterior* infinity – the infinitely large such as some believe the universe to be), but also, as Zeno apparently assumed, for something that decreases without limit (an *interior* infinity – the infinitely small such as mathematical notions about the points on a line). Surely, such a concept lies beyond the limits of ponderable comprehension and defies intuitive understanding. The truth is that we are able to deal with the notion of infinity only through non-intuitive symbolic manipulation.

Put another way, infinity apparently comes in two versions: 1) that which increases (extension) without limit and 2) that which decreases without limit aka the infinitesimal. The first should be replaced by the sense of endlessness, which can be a simple as traveling in a circle – the circumference is finite while the extent of travel around the circle is unlimited (unbounded). The second should be considered as the modern mathematical concept of limit such as that used in the

---

[113] David Bohm, *The Undivided Universe*, (Routledge, New York, 1993), 38.

fundamental theorem of differential calculus. Approaching zero is not zero.

Without doubt, it would be better if most of us just did not use the word infinite and find a less imponderable word (interminable, limitless, really big or small?) because what appears to be extension without limit could be just going around in circles.

The Void

"Whoever speaks of absolute space uses a word devoid of meaning. This is a truth that has been long proclaimed by all who have reflected on the question, but one which we are too often inclined to forget [Emphasis added]."[114]

"It will be seen that though geometry is not an experimental science, it is a science born in connexion with experience; that we have created the space it studies ... We have chosen the most convenient space, but experience guided our choice. As the choice was unconscious, it appears to be imposed upon us. Some say that it is imposed by experience, and others that we are born with our space ready-made [Emphasis added]."[115]

"To be material is to have a spatial position. The number 2, for example, has no spatial position. Spatially contiguous objects represent one another. We do not know what the objects are if we know merely that they are in space. We understand space only through the drive of the objects in space; otherwise we have no idea what space is [Emphasis added]." [116]

"Let us then see what we are thinking about when we speak of 'Nothing.' To represent 'Nothing,' we must either imagine it or conceive it. Let us examine what this image or this idea may be

---

[114] Poincaré 1918, 93.

[115] Ibid., 115.

[116] Kurt Gödel Quoted in *A Logical Journey*, Hao Wang (Cambridge MA: MIT Press, 1996), 292.

[Emphasis added]." [117]

The first quote questions the validity of absolute or physical space, while the second quote offers a choice between learned space and instinctive space. Herein, it is proposed that we learn the concept of space at an early age, and it is deeply ingrained such that it has the appearance of being intuitive. That is, because our senses limit us to a three-dimensional world, we assume that we are embedded within a three-dimensional space, although there is no substantive evidence to support this belief. To be sure, rooting out this notion of three-dimensional space, or any other, ingrained notion is difficult even when directly confronted with a viable alternative.

The third quote suggests that it is only through positional relations between objects that we build our notions about space, and, as the fourth quote states, visualizing nothing is not so easy and perhaps all we can do is define it.

Before continuing, I would like to list a few quotes that I gleaned from a well-known online philosophy of science forum about the believed nature of space:

"We know that space has properties, so it is real. One particular property prevents light from propagating at infinite speed [Emphasis added]." - Does this person really believe that it is a property of space that restricts the speed of light? Also, I would like to see a list of the properties of space.

"Physical space is then a container of objects, shared and common to all observers. This space only differs for each observer by their respective metrics." - So, for this person space is a box – then where are its boundaries and what is this box made of?

"It's better, I think, to view the relationship between matter/energy and space as one of coinciding or coincidence: particles or fields of matter or energy are said to coincide with parts of space. Rather than

---

[117] Bergson 1911, 293.

just a passive container, space has the power to structure matter in three and only three dimensions." – For this person, space is the source of three-dimensional matter!

I can only say that these three opinions by academic philosophers seem somewhat naïve, and I find their notions disturbing for their lack of critical thought and insight. I suspect that these quotes are partly the result of misinterpreting the theory of general relativity with the mistaken belief that there is a physical relation between space and gravity (this common mistake is discussed in the next section – *General Relativity and the Void*). Anyway, to continue...

If we remove all matter, what is left? Put simply, without matter there is nothing. As implied in the third quote- at the beginning of this section – the term space only has meaning within the context of matter. It is from the relational intervals between substantive objects that we build our notion of space. Material objects possess physical reality and a relational position, and every reference to an object requires other objects in order to establish these positional relationships.

As the fourth quote by Bergson reminds us, intuitively we want space to be *something* because we cannot directly experience nothing and must imagine it through thought.

After some thought, it seems apparent that without matter, there is only the Void, and therefore it is incorrect to speak of *empty space*, since empty implies a container of some sort. The Void – being nothing – has no dimensions, no extension; it is total absence. As nothing, the Void cannot be the cause of anything. If this is so, what then is the cause of our belief in three-dimensional space? This question will be answered later.

{**Side Comment on Giordano Bruno:** In the year 1600, Giordano Bruno, a Dominican Friar, was burned at the stake partly for his views on cosmology and partly on his lack of piety (to put it mildly). Bruno was more a philosopher than a scientist, but among other things, he heretically proposed that the stars were actually distant suns and that the universe contained innumerable inhabited worlds! For these and other beliefs, he was condemned by the Inquisition.

To see why this happened, it is helpful to take a brief look at some of his ideas that were published over the 20-year period prior to his demise. Although, some or even most of his ideas were not original, his manner of presentation was original. Bruno postulated that other than the stuff of the universe, there existed only the Void. For example, here are a few of quotes from one of his better-known works:[118]

"Sixthly, if we posit a finite world, it is impossible to escape acceptance of the void, if void is that which containeth naught."

"Seventhly, this space in which is our world would without it be indeed a void, since where the world is not, there we must infer a void."

Additionally, Bruno conjectured that earth is just one of an infinite number of worlds and he differentiated between the planets – that reflected light and the suns – that produce light.

"...it is shown that although some bodies are luminous and hot of their own nature, yet it doth not follow that the sun illumineth the sun and the earth illumineth herself, or that water doth illumine itself. But light proceedeth always from the opposed star..."

"...but those suns, bodies in which fire doth predominate, move differently to the earths in which water predominateth; thus may be understood whence is derived the light diffused by stars, of which some glow of themselves and others by reflection."

He also believed that many other earths exist that have life, and the glory of God has been made manifest in these worlds as well.

"Ninthly, since the famous order of the elements is seen to be vain, the nature is deduced of these sensible compound bodies which as so many animals and worlds are in that spacious field which is the air or the heaven, or the void, in which are all those worlds which contain animals and inhabitants no less than can our own earth, since those worlds have no less virtue nor a nature different from

---

[118] Giordano Bruno. *On the Infinite Universe and Worlds*, Venice, 1584.

that of our earth."

"...Thus, is the excellence of God magnified and the greatness of his kingdom made manifest; he is glorified not in one, but in countless suns; not in a single earth, a single world, but in a thousand thousand, I say in an infinity of worlds."

That is quite a list and after making the mistake of moving back to Venice, he was accused of heresy, turned over to the Roman Inquisition, and taken to Rome. There, for several years, he remained in a Roman jail where every now and then he was brought out for questioning and then returned. Exactly what charges were brought against Bruno is unknown but eventually, in 1600, he was finally condemned and burned at the stake at a piazza where now stands his statue.}[119]

General Relativity and the Void

The commonly accepted interpretation of general relativity states that space is bent and distorted under the gravitational influence of large bodies (planets, stars, etc.). This notion that space is curved is based on an interpretation of the mathematics of general relativity that tacitly assumes the existence of absolute space. That is, the mathematics of general relativity can be interpreted such that space is absolute in the sense that space is something that can be bent or otherwise distorted. Then according to this common interpretation of the equations of general relativity, gravitational forces either cause the assumed three-dimensional space to bend into another dimension or there is stress without bending.

Traditional discussions about general relativity often use models such as that of a bowling ball rolling around on a trampoline. This is supposedly analogous to the impact that large masses (planets, stars, and galaxies) have on three-

---

[119] The statue is in the Roman piazza *Campo de' Fiori*.

dimensional space. In this model, the two-dimensional surface of a trampoline is distorted into a third dimension.

Since it is best to go right to the source, here is what Einstein had to say about general relativity back in 1915 when it was first published:[120]

"We introduce in a space, which is free from Gravitation-field, a Galilean Co-ordinate System K(x, y, z, t) and also, another system K'(x', y', z', t') rotating uniformly relative to K. The origin of both the systems as well as their Z-axes might continue to coincide. We will show that for a spacetime measurement in the system K', the above established rules for the physical significance of time and space cannot be maintained [Emphasis added]."[121]

"Because among all substitutions there are, in every case, contained those, which correspond to all relative motions of the co-ordinate system (in three dimensions). This condition of general co-variance, which takes away the last remnants of physical objectivity from space and time, is a natural requirement, as seen from the following considerations. All our well-substantiated spacetime propositions amount to the determination of space-time coincidences. If, for example, the event consisted in the motion of material points, then, for this last case, nothing else is really observable except the encounters between two or more of these material points. The results of our measurements are nothing else than well-proved theorems about such coincidences of material points, of our measuring rods with other material points, coincidences between the hands of a clock, dial-marks and point-events occurring at the same position and at the same time [Emphasis added]."[122]

Both of these quotes clearly state Einstein's position on the reality of space and time. Without troubling about the K and

---

[120] In his later years, Einstein made the comment that after the mathematicians had taken hold of his theories, he no longer recognized them.

[121] A. Einstein, The Foundation of the Generalised Theory of Relativity (1915): https://en.wikisource.org/wiki/The_Foundation_of_the_Generalised_Theory_of_Relativity#C._The_Theory_of_the_Gravitation-Field, 5.

[122] Ibid, 6.

K' stuff in the first sentence, the first quote clearly states that maintaining any notion of a physical spacetime is untenable, while the underlined sentence in the second quote clearly states that the removal of the last remnants of physical objectivity from space and time means that both space and time are mental constructions and computational conveniences. These notions are both compatible and supportive of the concept of the Void and the non-reality of time.

That is, as presented by Einstein, and in agreement with the Void, there is no space to distort. The stress between the large bodies is a gravitational struggle, which only gives the appearance of a distortion of space. The physical effects of this apparent distortion result from *gravitationally defined trajectories* that behave exactly as curved space would, if space were something that could be curved.

In the next section of this original paper, Einstein goes on to mention that: *"...it is clear, that from the physical standpoint the quantities gab indices are to be looked upon as magnitudes which describe the gravitation-field with reference to the chosen system of axes [Emphasis added]."123*

This quote clearly states that the geodesics (the shortest path between two points on a sphere or in this case the shortest distance between any two extraterrestrial bodies), are proportional to the tensor quantities g_{ab}, which represent apparent curvature from *gravitational* effects and not space curvature.

That is, the apparent curvature that is generally attributed to space, is instead the curvilinear path dictated by gravity and the relative movement of bodies. For example, interplanetary probes do not follow the curvature of space; instead, they follow complex trajectories defined by gravitational fields (plus other factors, such as the relative motion of these massive bodies).

---

[123] Ibid, 7. (The *gab* are components of the basic metric tensor.)

Mathematical equations stand alone as syntactical statements (formal relations) that are devoid of semantics (content). These equations are always subject to multiple possibilities for interpretative content.[124] Agreement between the equations and the theory is necessary, but not sufficient. To illustrate this point, here is a quote by Henri Poincaré; *"Mathematicians do not study objects, but the relations between objects; to them it is a matter of indifference if these objects are replaced by others, provided that the relations do not change. Matter does not engage their attention; they are interested by form alone."*125

Put another way, the mathematics of general relativity, as with all mathematical equations, only describes behavior and form, not content, and through the introduction of content these equations are subject to different interpretations. Semantic free equations are mute when it comes to interpretations since mathematics is concerned with relations between conceptual structures, not with interpretive context. That is, they are concepts about structure, or they determine structures. This brings to mind a famous quote attributed to Einstein that: *"In so far as the structures of mathematics refer to reality, they are not certain, and in so far as they are certain, they do not refer to reality."*

As another example of how mathematical equations can be misinterpreted, the equations for interplanetary geodesics are often erroneously interpreted as curvature of space when more correctly, they should be interpreted as gravitational stress between massive bodies such as planets, stars, and galaxies.

When considering fundamentals, it is necessary to avoid thinking in terms of space and time as having some kind of physical existence. That is, the notions of absolute space and

---

[124] For example, the Inverse Square Law equations have the same form for both Newton's gravitation and for Coulomb's Law such as the mass of two bodies for Newton's gravitation or the intensity of two electro-static charges for Coulomb's Law.

[125] Poincaré (1952), 20.

time are an obstacle to clear fundamental thinking about the interactions of physical bodies and the structure of the universe on the broadest scale.[126]

It is unfortunate that after more than one hundred years, references to physical spacetime are still to be found in technical papers – although as originally put forward –Einstein's theories of relativity clearly rejected such notions.

Dimensionality

After the previous discussions, it may seem strange that the next topic should be dimensionality. Commonly, we think of dimensions as dimensions of space, but that is just because that is what we call them, and we are accustomed to thinking that way. To discuss *spatial* dimensions within the context of the Void is inappropriate. On the other hand, inasmuch as we commonly experience three dimensions, dimensionality is part of our reality.

It appears that many, if not most, arguments in favor of a three-dimensional universe simply assume that this is the correct number of dimensions and then proceed to defend this choice. In most of these arguments, three-dimensional space seems based on the assumption that, as discussed in the prior section on the Void, there must be some kind of container or at least an entity capable of exhibiting some number of dimensions. It seems that we naturally like things tidy and in boxes – in this case a three-dimensional box. However, since the Void is total absence, the Void has no dimensionality. If that is so, then why do we perceive a three-dimensional world?

---

[126] The Hamiltonian operator representing both potential and kinetic energy is a classic example of how a contextless mathematical equations can have application in multiple domains of physics.

The simple answer is that we experience a three-dimensional world because our senses are apparently only capable of building up a perceptual space of no more than three dimensions – beyond three dimensions we lose our intuition and ability to visualize objects of more than three dimensions. Our visual space is based on two-dimensional perception, which is processed through accommodation and correlation into three-dimensional images that are then commonly interpreted as objects in three-dimensional space.

This notion of three-dimensional space is also supported by our tactile space, which is constructed from our ability to feel objects only in three dimensions. In this way, the idea of three-dimensional space is built up through the limitation of our senses and the lack of mental capability to intuit or visualize beyond three dimensions.

Of course, these limitations do not preclude the possibility of hidden structure.

Hidden Structure Model

First, this model is not a representation of space, but is a geometric exercise that illustrates how, even in a simple model, structure can be hidden.

Consider a loop such as a circle or any closed figure that does not cross over itself (like *0,* not like *8*). This type of loop allows a moving point unrestricted movement, such that a moving point will eventually return to where it started. For example, visualize a coffee cup with a bug walking around the rim, this bug has unrestricted movement, but only in one dimension. Next, consider the surface of an inflated balloon. If a bug is walking on the surface of the balloon, the bug will have unrestricted movement in two dimensions.

However, in both examples, unseen from the bug's point of view, there is hidden structure. The one-dimensional loop (or coffee cup rim) is embedded on a two-dimensional surface.

The two-dimensional surface of the balloon also has a hidden structure, since the surface of the balloon is a two-dimensional surface embedded on a three-dimensional sphere.

Surface geometry deals with the outer measurement (cup rim, balloon surface), which gives the appearance of being independent of the geometry in which the surface is embedded, but the fact that there is unrestricted movement in both models relies on hidden structure. It should be noted that these examples may be extended into more dimensions, and the hidden structure requirement remains.[127]

Under certain circumstances, there are hints or clues that reveal possibilities about this hidden structure. For example, unusual occurrences, sightings, and what might be called non-hallucinatory visions are in some cases possibly a result of hidden structure. For example, there is the occurrence where one sometimes has the strange feeling of being watched and on looking around sees that this is so. Another example is the phenomenon called synchronicity. This term, coined by the eminent psychologist Carl Jung (German/Swiss, 1875-1951), refers to the synchronized occurrence of seemingly unconnected events.

Dabbling in Higher Dimensions

"Quantum physics in particular seems to indicate that physical reality is something still more different from appearances than even the 4-dimensional Minkowskian world. T. Kaluza's fifth dimension

---

[127] In the technical and philosophical literature, seeing this hidden structure is sometimes referred to as *God's view*, with the obvious implication that, like the bug, such hidden structure is unvisible to us but is observable only by Deity. On a personal note, some years back, I emailed the well-known Stanford physicist Leonard Susskind about hidden structure and his polite response was that I should consider it as analogous to the scaffolding used in constructing buildings. A nice response, but I cannot help feeling that there is more to hidden structure than that.

points in the same way."[128]

In 1909, the physicist Gunnar Nordstrom (Finnish, 1881-1923) speculated that the universe might include dimensionality higher than the three dimensions we normally experience. Unfortunately, Nordstrom published his ideas in an obscure journal and, consequently, his ideas received little attention and were soon forgotten. Following Nordstrom, there are the better-known efforts of Theodor Kaluza (German mathematician, 1885-1954) who first published on extra dimensions in 1919.

Later, in 1925, Kaluza collaborated with Oscar Klein (Swedish physicist, 1894-1977), to propose the use of higher dimensions as a way of building a unified theory of gravity and electromagnetism. Not much happened with this idea, but some years later, in 1943, Albert Einstein and Wolfgang Pauli (Austrian-Swiss physicist, 1900-1958) published a co-authored paper that cautiously approved of the Kaluza-Klein arguments.[129] Following that, in 1956, Pauli added, to a revision of his book, *The Theory of Relativity*, an extended supplemental note about the Kaluza-Klein solutions. In this supplement, he stated that although more work was need, this direction seemed on the right track.[130]

Unfortunately, the ideas put forward in the 1943 article, and the 1956 supplemental note, did not receive much attention. Nevertheless, these are examples where the concept of additional dimensionality was taken seriously, and in this last example by two of the most prominent physicists of the 20th century.

---

[128] Kurt Gödel, *Collected Works Volume III*, Eds. Feferman, Solomon (Oxford: Oxford University Press, 1995), 240.

[129] A. Einstein, and W. Pauli, "On the Non-existence of Regular Stationary Solutions of Relativistic Field Equations," *Annals of Mathematics*, Vol. 44, No. 2 (1943): 131.

[130] Pauli, 1981, 224.

As it turned out, all these ideas were essentially dormant until the 1980s, when *String Theory*, which later morphed into *M-theory*, became all the rage. Both theories relied on higher dimensions (10 for string theory and 11 for M-theory), although in these cases the extra dimensions are neatly compacted to the microscopic level to ensure that they remain out of sight. By *shrinking* the extra dimensions to the very small (Planck scale) these theories avoid both mathematical anomalies as well as not being required to explain why we do not experience the extra dimensions. However, after a period of lofty expectations, it became apparent that both String Theory and M-Theory were failing to live up to expectations. Subsequently, both have diminished in both luster and popularity.

The point is that interest in extra dimensions is not new, and it seems that interest has been growing over the last few years, possibly inspired by String Theory, such that dimensionality is again receiving serious attention. For example, in 1999, Lisa Randall (American physicist, 1962-) and Raman Sundrum (Indian-American physicist, 1964-) wrote a paper proposing that maybe all extra dimensions were not completely compacted,[131] and later Randall expanded on this notion in a book.[132] Another recent addition to the theme of extra dimensions is a book, published in 2006, by the physicist Paul Wesson (Canadian, 1949-2015).[133] These recent efforts favor extended extra dimensions rather than the rolled up compacted variety.

---

[131] Lisa Randall and Raman Sundrum, "An Alternative to Compactification." (1999) http://arxiv.org/pdf/hep-th/9906064.pdf.

[132] Lisa Randall, *Warped Passages* (New York, Ecco, 2005).

[133] Paul Wesson, Five-Dimensional Physics (London: World Scientific Publishing, 2006).

Part 1 Recapitulation

In some ways, conceptualizing time has always been more challenging and problematic than space. In the case of space, we have sensual perception to assist us. We may not see space, but we do see that there is a separation between objects, which includes both the separation of small objects nearby and the separation of heavenly bodies. This separation, the gaps between visible objects, is what we commonly call space.

On the other hand, time, a thing not directly perceived in any physical way by our sense perception, is an abstract mental construction that we use to explain the separation of events; in a manner, somewhat analogous to the way we use space to explain the separation of objects. The obvious difference is that the separation of objects is directly observed, while our sensation of time is not directly observed and is variable to the extreme. The only way we visualize time is geometrically as a series of unequal linear durations such as an unequal dashed line (-- —— - - - —).

Although humans are primarily guided by common time, remnants of natural time still lurk in the background and continue to influence human behavior. For example, there still seems to be a certain allure and fascination with the full moon. Although most of us no longer associate the moon with lunacy, perhaps there is a thread of truth about the behavioral influence of a full moon.[134]

It is difficult to say which is easier to give up, common time or common space. The various arguments for the existence of physical space are different and yet similar to those for time. They differ in content but are similar in that they tend to take

---

[134] Some years ago, a friend, who was an emergency center physician, told me that there seemed a definite connection between the full moon and activity at the emergency center. Perhaps various law enforcement agencies could corroborate this observation.

an anthropocentric view of the way things are in the universe. That is, there is a tendency to treat natural phenomenon as somehow connected to humans, as if human observation and interpretation of natural phenomena is a requirement for the existence of the natural world. For example, we seem unable to avoid certain anthropic references and use gender specific names in reference to both natural objects and human constructions. As another example, with exceptions such as English, there are other languages take this tendency to the extreme by gendering nouns.

An unforeseen consequence of Minkowski's unification of space and time into the concept of spacetime was that once united, they stand or fall together. However, even before this unity, time was never able to stand completely on its own without the support of physical events. It is only through events (change, motion) that even natural time has ever had any meaning.

Even though the theories of relativity eliminated the notions of time and space as something physical, I still see scientific and philosophical articles that refer to *physical time* and/or *physical space*. I can only assume that many people cannot imagine that without matter-energy content, there is only the Void, and that time only has meaning as a linear metric used to separate events, and that this measurement can only be performed by comparing one physical process (for example, a moving train) relative to another physical process (such as a clock).

Space and time are media for the ordering of the myriad affairs of humans. It is within these two concepts that we organize our lives such that we have come to believe that one is some universal container, and the other is also a physical reality that provides sequence, continuity, and a certain direction. We intuitively need the container (space) because otherwise we would feel exposed and vulnerable, and we need time as the measure of an order that flows only in one direction.

Questions about space and time have their origins in the ancient past and it is reasonable to believe that these questions will continue to be discussed well into the future. There are always individuals either unwilling or unable to put aside old ideas and adopt new ones. Realistically, there is no doubt that even with the notions of time and space that have been presented here, common time and space will not go away and will remain a part of our comfort zone.

In any case, it was not the purpose here to get rid of common space and time. Both serve useful purposes within our everyday lives. It is only when contemplating deeper issues that the true nature of space and time must be considered, and common notions of space and time are relegated to their proper status. For example, when time and space are considered in the context of fundamental science, philosophy, or theology the true nature of time and space needs to be maintained.

Part 2: Perception and Reality

Both perception and reality are vast and complex topics and for that reason there is no pretense of this work being anything like a complete or a definitive exposition of either perception or reality, but rather one, which will provide the reader with some novel notions about perception and reality, while being interesting and (hopefully) somewhat entertaining.

To accomplish this, Part 2 takes a quick look at common perception and common reality then both are reconsidered in more depth to show that they are complementary such that both influence the other and are, in fact, inextricably connected.

Considering Perception and Reality

Common Perception

Although a constant companion while awake, the process of perception is mostly ignored. Even now while reading these words, the process of perception, although probably not noticed, is at work. Only in dreamless sleep or a state of anesthetized unconsciousness do we escape this constant companion. Not only is perception persistent, but it is also our window into the external world.

Common perception, which might also be called *common sense* perception, is something we use in a manner like the way we use technology. Although we use technology on an everyday basis, for the most part, we only have *operational* knowledge of this technology. Operational knowledge is like following a recipe in that recipes provide necessary steps for usage without any explanations of the whys and wherefores. That is, we know how to use technology, but have little or no *functional* knowledge of technology. For example, the use of personal communication devices (aka cell phones) is common to the point of being universal, but very few users have even the slightest idea of how cellular communication functions. The same is true of perception, our perception is always active and in use, but our awareness of the process of perception is for the most part hidden or simply ignored.

During normal daily activities, we do not consider or question the details of perception, we simply accept what we perceive and let it go at that. Of course, this is very practical, since if we stopped to examine the details of the flood of received sensations, we would quickly become overloaded and find ourselves incapable of normal functioning. Normal activity requires that we not spend too much time analyzing our perceptions.

Perception as commonly experienced assumes the reality of the perceived. We intuitively feel that perceived objects have an immanent existence of their own just as we feel the reality of our own existence. What we usually do not consider is that it is through perception that we are able to build up our notions of reality and confidence in the reality of the world.

{**Side Comment on Sensation versus Perception:** Here is an example to help illustrate the difference between sensation and perception. When a young child or infant first encounters lemonade, they have a sensation of the taste of the lemonade as a simple undifferentiated stimulus. With experience and more maturity, the lemonade becomes a compound sensation, a complete perception. That is, as children gain experience, they begin to differentiate the lemonade into components such as sour, sweet, and cool. The initial sensation of lemonade develops into the more complex and complete perception of lemonade.}

Common Reality

Herein, the definition of common reality is the reality of common or everyday perception. Although somewhat circular and not a particularly useful definition, it is more or less accurate. Put another way, our reality is centered on ourselves, that is, our complete self, both physical and mental (the psychophysical self). Everything we experience is a projection onto and into ourselves; and it is through the reception of these projections that we build our reality and total view of the external world. For example, while writing this sentence, there is a sense of contact with the computer keyboard, which together with the computer and the immediate surroundings is considered real.

These things are *real* because they are persistent and familiar. That is, confidence in the reality of *ourselves* and the *not-ourselves* is built up through perceptual persistence of objects and the familiarity of recurring events. It is our confidence in this common reality that carries forward into the future with expectations of consistency that are essential for our sense of well-being and normalcy.

Commonly, we usually do not differentiate between the words real and reality since in common reality they essentially coincide, and it is only when we dig deeper into these concepts of what is real and what is reality, do we find that they are indeed very different notions.

Concept Space

"Nothing is presupposed save the existence of things in their inexhaustible multiplicity, and the power of the mind to select from this wealth of particular existences those features that are common to several of them."[135]

"...concepts do not gain their truth by being copies of realities presented in themselves, but by expressing ideal orders by which the connection of experiences is established and guaranteed."[136]

"Concepts are there, but not in any definite place. They...form the 'conceptual space'."[137]

Concept space is the mental space of concepts constructed through selective analysis and classification of our perceptual encounters with objects and events. Although the process of forming concepts and use of concepts goes largely unnoticed, this concept space is indispensable for normal functioning.

Most dictionary definitions define concepts as ideas or images that are abstractions or mental impressions. Such definitions are correct, as far as they go, but they are also incomplete and fail to do justice to the importance of concepts. More than simply mental impressions or representations, concepts are abstract generalizations, and associative correlations that provide an essential capability for dealing with the continuous flow of complex sensations and perceptual multiplicity.

---

[135] Ernst Cassirer (German philosopher, 1874 – 1945), *Substance and Function*, (New York: Dover Publications, 1953), 4.

[136] Ibid, 319.

[137] Kurt Gödel [Austrian-American logician, mathematician, and philosopher, 1906 – 1978] quoted in: Palle Yourgau: *A World Without Time*, (Cambridge, MA: Basic Books, 2005), 175.

Concepts are both efficacious and essential elements of our capacity to handle the constant bombardment of massive amounts of sense data. Through a process of abstraction, summarization, and classification, concepts facilitate our ability to deal with copious amounts of data as generalized patterns, templates, and abbreviations.

Put simply, concepts are indispensable short cuts for rapid recall, decision-making, and just plain normal functioning. To be sure, concepts play a necessary role in essentially all aspects of our lives, but perhaps the primary use of concepts is to facilitate accommodation, assimilation, and classification of phenomenal experience.

A concept is an association with the vanished content of past objects and events, which lack the exactness of these past objects and events, is often only a blurred remembrance. A concept is not that which is perceived (the percept), and it is not possible to effectively visualize a concept. That is, whether a concept has only one referent or a multitude of referents, the concept itself is not one of the referents.[138] That is, the concept itself is not a member of the concept's set of referents and cannot be instantiated or exemplified. Attempting to visualize a concept will only result in the image of a known particular or referent.[139]

Concepts are material for reflection and contemplation. New concepts are synthesized through recognition and association of *similar-differences* and *different-similarities*.[140] They mediate between thought and language such that they assist in the creation of new levels of knowledge. Through

---

138 A referent as defined herein can be a physical object, an event, a fictitious thing, or a thought (such as an emotion).

139 This is also true of individual concepts such as the concept of a pet dog is not the image of the dog but a compound collection of images.

140 For example, tree leaves have similar purpose, but difference structures and roses have different colors, but similar structures.

concepts emerge levels of meaning external to and independent of the sources of this extracted knowledge. That is, concepts, being removed from place and time, are available for free associations such that they assist in the building of new inferences and new content. Concepts create categories *of* and *for* understanding.

Concepts are the connections between the abstract and the concrete world of sensations – they increase our ability to handle physical reality as well as the ability for rational creative thought. Concepts evolve as experience evolves – every thought, whether practical or whimsical or anything else, is full of conceptual content.

Concepts can be divided into two basic categories of *thing-concepts* and *event-concepts*. Thing-concepts are somewhat self-explanatory, while event-concepts, as the name implies, are associated with actions – especially repetitive actions – and are basic to both humans and the higher animals. Examples of this type include such everyday activities as writing and driving an auto. However, they also have a higher level that includes processes such as interactions within family and non-familial social interactions. Put another way, event-concepts include most of what we do every day and are learned through our interactions within our environment.

Types of Concepts

Although there are many ways to classify concepts, the two categories just mentioned can be sub-divided into three sub-categories or types that are sufficient to include most, if not all concepts. These three sub-categories will be called: *attributive*, *generic*, and *individual*.

Attributive: These concepts are probably the type most recognized and represent the simplest type of concept. As the name implies, this type of concept represents the phenomenal attributes or properties of objects (the referents) that are used to define and identify the object. Examples of attributive concepts are many and include such notions as *red, furry, biped,* and so on.

Open any dictionary and select the definition of any object. It is certain that the definition chosen will refer to attributes of the object and that the definition assumes a basic understanding of these attributes. Attributive concepts, of course, are not restricted to properties of objects, but also include properties of events, such as *sunny, cold, outdoor,* and so on.

It might be argued that attributive concepts are not true concepts, but simply attributes or characteristics. However, consider the concept of a particular color, since the concept itself cannot be visualized, our experience of a particular color is never the conceptual color, but rather an example of the conceptualized color. For example, when we consider the concept *blue,* we visualize a particular shade of blue, which is an exemplification of the *blue* concept. That is, the *blue* concept is the standard by which we recognize and associate all shades of blue.

It is through adjectival modification that we extend concepts by concatenating them with other concepts. For example, the perceived blue might be *light-blue, red-blue,* or any other shade of blue. Put somewhat differently, when we say, "*This is blue,*" we have a relationship between a real object as perceived and an exemplification of the *blue* concept.

Generic: These concepts represent the aggregated properties of objects. These concepts are composite concepts and always lack the simplicity of attributive concepts and always have multiple referents. For example, the generic concept of *tree* is the aggregate of the attributive and generic concepts common to all trees. As the name implies, generic concepts are aggregated generalizations, and it is through these generalizations that we build conceptual hierarchies.[141]

Generic concepts form classification levels such that hierarchical relations exist between these levels. Concepts of a lower (more specific) level are inclusive within the higher (more general) levels, while the lower levels are exclusive of all other concepts included within the same level. For example, the *pine* concept is included in the more inclusive generic *tree* concept and is exclusive of all other types of trees – such as the *oak* concept, while the concept of *tree* is included in the more inclusive generic concept of *plant*, such that within the hierarchy, all pines are trees, and all trees are plants. On the other hand, the tree concept excludes all non-tree plants.

The hierarchy of generic concepts relies on either adding or deleting various attributes (or properties). Adding attributes leads to greater specificity, while deleting attributes leads to greater generality. At one extreme is the specificity of but a single exemplification, and at the other end of increased generality is a categorical apex such as the concept of *plant*.

This is not to say that all reach the same apex, but rather due to category scope and the complexity of conceptual thought each category eventually reaches an apex or terminating node. For example, *plant* – includes all plants - and *tree* – includes all trees.

Individual: As the name implies, this type of concept represents only a single thing or referent. It might seem strange to consider an individual object as a concept; however, when thought about in a certain way, such a notion, or should I say concept, is easy to accept. For example, each human individual exemplifies the attributive and generic concepts associated with all humans, plus the attributive and generic

---

[141] It should be noted that all events are conceptually generic. Every event is a complex set of objects and action, including the observer and the observed.

concepts that uniquely individualize this particular person. For example, an individual named Napoleon, which as the *Napoleon* concept is dependent on a multitude of concepts such as, *human, biped, Corsican, French Emperor, married to Josephine*, and so on.

Every person we know is an *individual* concept. Furthermore, each person knowing the same person has their own unique concept of this person that is customized based on relationship, past interactions, personality (of both persons), and so on. Individual concepts have no restrictions on referents except, as the name implies, individual concepts require that the individual thing (person, animal, object, event, or thought) have a uniquely discernable identity.

Non-real Concepts

What has been discussed above primarily concerns *thing-event* concepts, which reference *real* (physical) objects and events. However, in this section, three subtypes of *non-real* concepts are considered: *intimate, fictional*, and *relational*. Although most of what has been discussed also applies to non-real concepts, these concepts, which are examples of generic concepts, are different enough to warrant a separate discussion. Since all concepts are abstractions existing in concept space, it might seem strange to refer to non-real concepts. However, non-real concepts are those that have no direct connections with concrete objects or actual events. That is, the referents are unreal.

For example, the referents of *intimate* concepts exist only as mental constructions, however, the construction of these concepts relies, at least partially, on personal experience. For example, the emotion of sadness or the sensations of pain are learned through experiencing events that evoked an emotional and/or psychophysical response and it is the repetition of these experiences that leads to the construction of *non-real intimate* concepts such as *sadness* and *pain*.

Although, initially motivated by actual events, as these concepts mature, they have a tendency, except in the cases of extreme trauma, to become disassociated from these events and exist as purely mental notions. This type of concept includes not only emotional states but includes any thoughts that do not readily lend themselves to visualization or verbalization. For example, intimate concepts associated with highly emotional symbols such as those associated with religion or politics.

On the other hand, *fictional* concepts are non-real concepts that relate to imaginary objects or events. For example, the concepts of *unicorn* and *Captain Ahab* are fictional concepts. Although fictional, these concepts rely on real *thing-event* concepts. For example, the *unicorn* concept relies on the concepts of *horse*, *horn*, and so on, while the *Captain Ahab* concept relies on the concepts of *human*, *ship's captain*, and so on.

The third type of non-real concepts to be considered are the *relational* concepts. These concepts are about comparisons or associations between two or more things and are included within the broader concept of *order*. Even though there are many types of order, commonly, we tend to think that order has something to do with the systematic arrangement of physical objects or a particular way of doing things. Our conceptual understanding of order is not intuitive but is one of the first things we learn.

Starting in early childhood, parents and others teach us the concept of ordering things. First, just by lining things up and later as we develop language skills, our sense of ordering becomes more complex and concepts like *alphabetical order* and *numeric order* become familiar. The referents of relational concepts can be either objects, events, or pure thoughts (such as emotions), but the relational concepts themselves are context free. For example, the notions of *greater than, before-after, friend of, in-between,* and *happier than* are conceptual relations, which are contextless and available for free associations.[142]

Non-verbal Concepts

Apparently, there are those who consider linguistic capability essential for the possession of concepts and apparently believe that verbal (word-like) concepts are the only kind that can exist.

This ignores the fact that all pre-linguistic children (and non-human higher creatures) have what might be called an innate ability to construct concepts. That is, pre-linguistic humans and creatures capable of true learning have an innate ability to abstractly process perceptions by selectively extracting similarities (and differences) without overt conscious effort. And, in a manner similar to the acquisition of motor skills, this innate ability must be exercised to improve functionally such that, as the individual matures, this ability also matures, and concepts become more complex.

Also, we should consider whether there is a difference between human concepts and what some call *animal representations*. As far as we know or are able to determine, the difference between humans and the higher creatures is that only humans take this process of abstraction further with the contemplation of concept themselves. To say that animals, being non-linguistic, only have mere representations is just another example of anthropic thinking and the making of unwarranted assumptions about human uniqueness.

That is, associating the lack of linguistic capability with non-conceptual representational thought and linguistic capability with conceptual states seems suspect since there

---

[142] Mathematics is a system of relational concepts where the syntax is defined by axioms, definitions, and other rules of the game. For example, consider this simple equation: $1 + 1 = 2$. This equation incorporates a number of non-real (abstract) concepts such as the numerical concepts '1' and '2', as well as the concepts of addition ($+$) and equality or equivalence ($=$).

appears to be no physical or cognitive reason for this belief. That is, the assumption that non-linguistic humans and animals are only capable of non-conceptual states has no physical or mental basis and arbitrarily precludes any pictorial or other non-verbal concepts such as those associated with our olfactory sense. For example, *sweet* and *sour* are concepts learned well before acquiring verbal analogs of these sensations.

By necessity, these non-verbal concepts are abstractions or pictorial representations of audile, tactile, visual, olfactory, and motile sensations; however, regardless of the form of these concepts, they apparently serve their purpose. For example, both pre-linguistic humans and higher creatures are known to react to signals in a manner that indicates learned generalizations, and these generalizations indicate conceptual thought. For example, all animals immediately recognize others of their own kind, which implies that they have something like a concept of their species, and non-verbal infants definitely know how to communicate their basic needs.

We learn about objects and events before we learn language and it is from these early experiences that we build non-verbal concepts – image concepts such as that of a toy (nameless but recognizable) and non-image concepts such as discomfort or hunger – that are associative mental constructions of these objects and events, which, later, after the acquisition of language they become associated with words. But even then, we still maintain non-verbal concepts where either a corresponding word does not exist or is unknown to the person.

That is, denying non-verbal concepts ignores the fact that even humans with language skills do not actually give up all non-verbal thought. Thinking in non-linguistic images is not only common, but also essential because of the limitations of language when handling complexity or ineffability that invokes strong non-verbal emotions such as those associated with religion and other highly emotional human issues.

Concepts and Perception

Contrary to what some might think, concepts do indeed serve a dual purpose of both the practical and the abstractly theoretical. They serve the practical by providing readymade notions about the sensible world of everyday living and they serve the theoretical by providing readymade cognitive shortcuts. That is, the practical uses of concepts are inseparably connected to our perceptions, while the abstract concepts of pure thought are part of our imagination, desires, fantasies, and such.

Although essential for life as we live it, concepts lack the fullness of active perceptions. That is, concepts, regardless of importance, lack the vividness of our perceptual flow especially during these special moments where something extra is revealed about the natural world that can be experienced in no other way than firsthand.

However, just because concepts are never as fully complete as perceptions does not mean that they are static, on the contrary, as we experience more, existing concepts can and do change through assimilation and accommodation of new content. For example, as I matured from childhood to adulthood new experiences of different food types and methods of preparation modified my concepts of some foods in particular and all food in general.

Since perceptions are fleeting and never reproduced exactly, we must rely on concepts to 'rebuild' past occurrences through recall and the accommodation of new perceptions.

Summary

When we put it all together, we see that generic and individual concepts are truly multi-dimensional such that they form an n-dimensional space where n can be a very large, but finite number that increases with the complexity of the concepts and completeness of our knowledge. When we think of a complex thing, we do so via a set of concepts, which constitutes our knowledge of something both generically and specifically.

Put another way, concepts are building blocks for associations and classifications that enable us to assimilate and correlate the unfamiliar by comparison with the familiar – or put still another way, concepts provide the necessary components that enable us to quickly discern perceptually encountered similar-differences and different-similarities. It is how we are able to quickly recognize both normal perceptions and abnormal perceptions such as an unexpected hazard.

As mentioned previously, we know other humans by their generic concepts that are common to all humans and by the individual attributes possessed by only a single instantiation of the *human* concept. However, due to the limitations of human capabilities, conceptual associations of complex entities such as humans are always incomplete.

As discussed above, concept space is a necessary adjunct to human life wherein we have developed language skills that enable us to associate objects, events, and thoughts with word labels, which is a - sometimes quite subtle - lifelong process of determining, selecting, and classification of similarities and differences.

Concepts allow us to extrapolate beyond the given world of perception. They provide us with the indispensable tools and capabilities necessary to think at levels of complexity that would otherwise not be possible. Just as mathematical equations are capable of *compressing* complexity into a few symbols (that are themselves concepts), concepts compress information and knowledge into manageable pieces that might be called intellectual or cognitive abbreviations. Without doubt, it is difficult to overstate the importance and indispensability of concepts. Essentially everything we experience and contemplate involves concepts. These very words are concepts, the act of writing is a concept, and the thoughts expressed in speech and writing are concepts.

Reconsidering Perception and Reality

After briefly discussing some aspects of common perception and common reality, it is time to reconsider these notions in a more analytic manner and take a look beyond the common notions of perception and reality. As this process proceeds, we shall find that both are more complex and interconnected than normally considered. In fact, although both are discussed separately, it will be shown that perception and reality are concomitantly connected and that any discussion of one necessarily entails the other - either explicitly or implicitly.

Perception Reconsidered

"But even the places in which I find myself are never completely given to me; the things I see are things for me only under the condition that they always recede beyond their immediately given aspects. Thus, there is a paradox of immanence and transcendence in perception. Immanence, because the perceived object cannot be foreign to him who perceives; transcendence, because it always contains something more than what is actually given. And these two elements of perception are not, properly speaking, contradictory."[143]

"...the senses are the organs through which the live creature participates directly in the on-goings of the world about... In this participation, the varied wonder and splendor of this world are made actual..."[144]

"It is in reference to our own body that we locate exterior objects, and the only special relations of these objects that we can picture to ourselves are their relations with our body. It is our body that serves us, so to speak, as a system of axes of co-ordinates."[145]

The authors of the above quotes clearly recognized that perception is not merely our *window into the world* but is the fundamental source of knowledge and human experience, and that this *window* is, through our sense organs, centered around our bodies. Therefore, expanding our understanding of perception beyond the elementary (or common) level provides the opportunity for both increasing our self-understanding and enhancing our worldview.

---

[143] Maurice Merleau-Ponty, *The Primacy of Perception* (Evanston, IL: Northwestern University Press, 1964), 16.

[144] John Dewey (1934), *Art as Experience*, (NY, Penguin Books, 2005), 22.

[145] Henri Poincaré, *Science and Method*, Trans: Francis Maitland, (London: Thomas Nelson and Sons, 1918), 100.

As stated in the first quote, perception is immanent because it is presented to us in a very personal way and that which we perceive is believed to have a concrete reality. On the other hand, perception is transcendental because beyond the reality of that perceived, the thing-in-itself remains unknown and, perhaps, unknowable. That is, our perception is always incomplete; we only have phenomenological knowledge of objects and events. Even if we consider this knowledge empirical or scientific, we cannot pass beyond the phenomena except by speculative inference and extrapolation.

The notion that through perception we live reality as experienced is another way of stating the primacy of perception. Perception involves a transference of the physical percepts into the mental realm of ideas, concepts, and thoughts. When an object or event is committed to our perceptual storehouse the physical object is not directly changed, but its phenomenological image is integrated into the Mind where it is wedged in with a myriad of other mental occupants.

Contrary to other forms of life, where the focus is on sustenance, reproduction, and survival in an environment, human activities include contemplating the utility of things and events, as well as seeking meanings for these things and events. That is, the human mind reflects upon itself (self-reflexive) and ponders the world in a general way. As far as we know, or are capable of knowing, on this planet, only humans pursue endeavors that seek the function and meaning (or purpose) behind the phenomena.

Although there is, in general, consensus concerning our common perception of objects and events, there is always some differentiation as well. Each of us has a different understanding and interpretation of perceived objects and events that reflect our personal preferences, experiences, and general inclinations. For example, when I first see a novel table, without conscious effort, I immediately correlate this table with personal encounters of other tables. These prior encounters with other tables have developed through a process of synthesis and abstraction into a generic image of table, which might well be called a conceptual notion of *table-hood* and through this process each of us has a different notion of table-hood. For example, a furniture artisan will *see* a table in a more distinctive way than most of us.

Perception has common elements based on intersubjective agreements by society at large, as well as personal elements unique to the individual. Put another way, perception has an interpretive aspect that is unique for every person.

Even science is only an extension of what we do in our everyday lives. What makes science different is the use of formal procedures and rules to extend this synthesis and make it more precise. In everyday situations, affirmation of the table as a table is immediate and easily understood, while, for the objectives of scientific study, determination and affirmation are more formal and more metrically exact.

The perception of an object is individualized, such that each person has a personal version of the object perceived. To say that the astronomer sees the moon more accurately that a Bushman of the Kalahari is not a statement of value, but a statement about a technological viewpoint versus what might be called a natural or spiritual viewpoint.

Earliest Perceptions

"Pure sensations can only be realized in the earliest days of life. They are all but impossible to adults with memories and stores of associations acquired."[146]

"The child's notion of an external dynamic world grows up hand in hand with his notion of himself, and his attitudes toward this world are as genuinely parts of his perceptive process as are the fusion of

---

[146] William James (1890*), Principles of Psychology Vol. 2.* (New York: Henry Holt and Co., 1910), 7.

sensory content which results in what we have learned to call the percept."[147]

The infant human has no past and no knowledge of the future, instead, the infant lives in the world of animal innocence, where there are no regrets of the past and no anxieties for the future – both the newborn human and the other live creatures live only in the present., Unlike many live creatures who are precocial, with only a few exceptions such as the need for nourishment, humans are physically underdeveloped and have few innate intuitions such that for the most part we are born without intuition of the outside world or of ourselves. Put simply, the initial world of an infant is a world of sensations without interpretation or meaning. The initial movements of a newborn are uncontrolled, and visual acuity is limited due to lack of the ability to differentiate sensory data into individual objects and events. For the most part, the newborn infant experiences only meaningless sensations.

As infants develop during the first few months after birth, they begin building an extrospective image of the outside world while at the same time building an introspective image of self.[148] That is, the child's discovery and knowledge of the outer world is inextricably connected with its discovery and knowledge of its inner self. Put simply, as the child discovers the outer world, the child discovers itself. As the child develops, these sensations take on meaning through sensory interactions, repetition, memory, and the development of the ability to differentiate these sensations and establish perceptual order.

---

[147] James Pratt, 'Realism and Perception'; *The Journal of Philosophy, Psychology and Scientific Methods*, Vol. 16, No. 22 (Oct.23, 1919), pp. 596-603, URL: <http://www.jstor.org/stable/2940132>

[148] To be sure, prior to birth, there are prenatal sensations, which include, for example - the voices of parents.

Within this process of increasing recognition and organization there is an inward movement that assimilates knowledge of objects and an outward movement where objects are manipulated from various viewpoints. The external sensations are enfolded through perception, while the child unfolds their intentions through the manipulation of the objects until they have satisfied their intentions toward the object. For example, a child will observe an object from various positions to gain a mental picture of the object's shape, texture, and overall appearance. This is essentially a process of learning about an object through a process of perceptually enfolding the phenomenal properties of the object.

Through constant experimentation with their sensory functions, indispensably aided by memory, the child builds up a more controlled and consistent behavior, and through repetition, the child eventually acquires a certain sense of order from these initially chaotic and seemingly random sensations and classifies these sensations - in a non-verbal way. This order takes the form of conceptual patterns of identification and association. For example, among the first patterns recognized are the faces and voices of their parents.

Throughout these early months, the child is developing generic summaries of the observed properties of objects and events, which are patterns or templates for conceptualization. Later, as language skills develop, they have preexisting classifications (non-verbal concepts) with which they learn to match words. Initially, through mimicking sounds and later through associations of particular sounds with particular objects and events.

Language development requires the association of abstract sounds (spoken words) with physical objects and events. Once the child learns to make logical connections or inferences between words, events, and objects, they have taken a major step forward in their development. This process, facilitated by the previously acquired store of non-verbal conceptual images, enables the child to make connections between spoken words and these conceptual images. After learning to make these connections between objects, events, and words, the development of language and other skills advance rapidly.

Perception as a Process

"...in both sensation and perception, we perceive the fact as an immediately present outward reality, and this makes them differ from 'thought' and 'conception,' whose objects do not appear present in this immediate physical way."[149]

"For the only way in which objects could conceivably designate our future actions would be by holding up to view the objective effects of those actions; that is to say, presenting the prior environment as it will be when modified by our reactions upon it. Perception would then be anticipatory, prognostic; it would exhibit to us in advance the consequences of our possible actions. It would thereby facilitate a choice as respects them, since the act of appreciating in advance the consequences that are to accrue from incipient activities would surely affect our final action."[150]

The first quote reminds us that perception of the physical world is the source our conceptual thoughts about the objects and events perceived, while the second quote reminds us that what we perceive provides us with choices as to possible actions and possible consequences of those actions.

Although ongoing research has made considerable progress in understanding the process of perception, the exact mechanisms are still only slightly understood and, for the most part, remain a mystery. Consequently, all we can do is describe the process in broad strokes that lack specificity.

---

[149] James 1910, 2.

[150] John Dewey, "Perception and Organic Action," *The Journal of Philosophy, Psychology and Scientific Methods, Vol. 9, No. 24* (Nov.21, 1912), pp. 645-668. The Journal of Philosophy, Inc.; <//www.jstor.org/stable/2012697>.

Common perception, as previously discussed, appears to be a singular event that happens almost immediately. The rapid pace of perception compresses the process of perception such that it has the appearance of a monolithic event. However, even the most fleeting and trivial perception is a complex psychophysical process. Therefore, to assist in examining this process, we will *slow down* the flow of perception by breaking it into four parts.

1. **Raw Data:** This is the external sensation that starts the process of perception. The term *raw data,* although common within information technology, might be somewhat strange to others. It essentially refers to data that has not been processed by a computer or any other means. As used here, raw data is the initial stimulus or sensation, which activates a particular type of sense organ.

2. **Reception and Translation:** This is the acceptance and encoding of the raw data by the sense organs, each accordingly to its sensitivity, into a form compatible for psychophysical processing.

3. **Assimilation and Accommodation:** This is the psychophysical process of transforming data into information.

4. **Decision for Action:** What we do with the information. This action may or may not include a motor (physical) response, but it always has an internal (mental) response.

The Flow of Perception

The future merges into the past via the specious present. This merging smears out the successive durations, which have a certain depth, and connects current events with the recent past into what seems a continuous flow. The perceptual experience involves the past and a potential future, separated by the immediate or current duration [duration is discussed in more detail in Part 2].

The brain and other physical components provide the interface between the Mind and the senses. That is, the process of perception includes something that might be called a brain-event, which is neurological activity that is associated and concomitant with Mind but is not the same. In this sense, the physical part of perception serves as a nexus between the purely physical and the purely mental.

An immediate perception, such as reading, brings the past forward through the recollection of past events. For example, when reading, it is not the perception of the words that has our attention, but the meaning of the words. The flow of perception simultaneously invokes existing memories and creates new memories, while the immediately active stimulus (a sequence of words in the case of reading) is continuously changing. On the other hand, the past, which was once future potential that became momentarily active, and then became a passive memory, is made active again within the perceptual duration. This is the process of apperception where new perceptual content is assimilated with existing mental content.

As the future becomes the specious present, we respond with either a passive or an active response to the current stimulus. That is, the *passive* reaction to the current perception is without overt motor response, while the *active* reaction includes a motor response. In the case of normal or routine recognition, the reaction is mostly passive, and we quickly discard the perception, and move on to the next.

1. Raw Data

Whether it is a cool breeze on your face or the vision of starlight finishing a multi-million-year journey, both are examples of raw data that start the process of perception. The sources of these stimuli are as varied as there are things and events in the world, and they are the *messengers* of the external world. Whether it is molecules of air pushing against our tactile sense organs or an electromagnetic visual event, the sources of this raw data are independent of the perceiver.

Since we are all familiar with our senses, it will not be necessary to discuss each type individually, but simply to state that each is unique and only receptive to particular types of raw data.

2. Reception and Translation

Since the types of raw data that we intercept are different, and the sense organs are sensitive only to their own particular type of data, by necessity there are organic mechanisms that convert these distinct types of data into a form usable by our cognitive functions. Essentially, the sense organs perform an encoding (or enfolding) that converts each type of raw data into electro-chemical sense data formatted in a pattern capable of processing by the recipient's cognitive functions. That is, each sense organ interfaces with the psychophysical self via electro-chemical pathways to the physical cognitive components.

3. Assimilation and Accommodation

This is the most complex and least understood part of perception. As mentioned above, the seemingly unconnected random stimulation of the sense organs, as experienced by infants, is the beginning of the multilevel and complex process of sensory assimilation and accommodation. This process, beginning with prenatal life, matures as the person matures, and continues in one form or another, for a lifetime. As the infant grows through childhood and into maturity, this process continues such that ideas and images are acquired and modified at each stage of development.

As perceptual objects and events flow through our senses and into memory, they are subject to mental manipulation. In this manipulative process, patterns are recognized, sorted, and classified through recognition of *similar-differences* and *different-similarities*. This is a process of pattern recognition and is not just about images of objects but includes events as well. To be sure, events also have patterns and the recognition of content and patterns within events is just as necessary as the recognition of objects. Although no event is ever repeated exactly, there will always be similarities and differences. Just as with familiar objects, familiar events assist in creating our sense of normalcy and reality.

Because sense organs receive data continuously and in enormous amounts, it is necessary to make a quick determination of what part of this constant sensory bombardment can be ignored and what should be retained for possible action. The solution is to automatically filter out most received sensations and retain only those that are judged necessary or desirable.

As I write these words, my attention is on the page, the pen, the words, and my thoughts about what I am writing such that while writing these words, the surrounding room receives little or no attention, unless something out of the ordinary occurs. This filtering and ignoring of *background* sense data is the only way we can keep from being overwhelmed with sensation and it facilitates our ability to focus by reducing distractions.

The immediate processing of sensations involves determining whether the sensations contain familiar or unknown content. If nothing novel or unexpected is encountered, then the immediate moment is quickly accommodated, and we move on to the next. On the other hand, if the content is unknown or unexpected then assimilation and accommodation of this content requires additional processing before taking action.

4. Decision for Action

The last step in the process of perception is a decision or judgment for action. At this point, we know enough about the current sensations to make a choice about what action, if any, we will take. This step is not a prediction of the future but is a process of selection and consideration of choices for action.

As stated in the above quote by Dewey, what we perceive presents choices with different consequences and the process of choosing involves an appreciation of the consequences of the possible options. Our ability to make decisions quickly is facilitated by prior experience, the recall of past events, and the result of actions associated with prior events. On the other hand, when confronted with situations containing new or novel content, we must improvise based on best guess and any prior information that appears relevant. This requires extra effort and if a judgment is not quickly obtained, indecision and mistakes are likely such as when someone panics.

This might be called the *reflective* part of perception, where after the initial processes of perception (parts 1-3) have essentially completed, a final judgment or consideration is made. Fortunately, most of the judgments we make are routine and not very exciting. In fact, we make decisions all the time without conscious effort or even the awareness that a judgment has been made.

Of course, these four parts of perception happen very rapidly and usually without overt consideration. This is especially true when encountering everyday common events.

{Side Comment on Information versus Knowledge: There is a hierarchical progression where data antecedes information and, in turn, information antecedes knowledge. That is, we progress from sense data to information, and from information to knowledge such that the acquisition of knowledge is always a comparison or correlation of new information with existing knowledge. Put simply, sense data becomes information when it is translated into mental content, and information becomes knowledge when it is comprehended and integrated with preexisting mental content. Sometimes information and knowledge, as well as data, are used as synonyms. This, of course, is a mistake, although these words are somewhat connected, they are not identical in meaning but form a cognitive hierarchy.}

Reality Reconsidered

"Physical reality, then, is the realm of actual and possible sense perceptions. The concretely physical is just that which yields and sustains perceptual experiences."[151]

"The belief in physical reality is really, in a final analysis, belief in a public realm of experience, accessible to other percipients of like nature with oneself. This belief, therefore, rests on the recognition of a social realm of beings with the same perceptual and rational powers."[152]

"The aim of the whole process seems to be to reach as complete and united a conception of reality as possible, a conception wherein the greatest fullness of data shall be combined with the greatest simplicity of conception. The effort of consciousness seems to be to combine the greatest richness of content with the greatest definiteness of organization."[153]

As the above quotes state, our reality has both personal and collective aspects. That is, reality is based on personal experience of the world as presented and interpreted by our senses and partly on community such as through intersubjective agreements. Since perception is the primary source of our reality, understanding perception is necessary if we are to understand our reality. As mentioned, experience of the external world through accumulation, accommodation, and assimilation of sensations forms the basis for reality. Experiences of external sensations are correlated and enhanced through mental processes such as extrapolation, induction, deduction,

---

[151] J. A Leighton, "Perception and Physical Reality." The Philosophical Review, Vol. 19, No. 1 (Jan. 1910), pp. 1-21, Published by: Duke University Press on behalf of Philosophical Review, URL: http://www.jstor.org/stable/2177636, 10.

152 Ibid, 16.

153 Josiah Royce, The Religious Aspects of Philosophy (Boston and New York: Houghton Mifflin Company, 1885), 357.

and so on. Although not completely reliable, these processes are the only way we gain knowledge of the external world.

Besides the word reality, we also use the word *actual* to designate what we feel is real. However, the word actual has a somewhat different usage, as it is often used to emphasis current existence, a recent occurrence, or an immediate presence. For example, statements such as *I was actually there.* indicate an emphasis on being personally present.

Consider also the words *reality* and *real*. Reality is the world of phenomena and our perception, real is the world behind the phenomena and perception. Reality is the thing as perceived, real is the unknown and, possibly, unknowable things-in-themselves.

Types of Reality

Within technical literature, of the type that discusses such things, we encounter a plethora of realities and realisms. Each was conceived to meet a perceived need, and each type of reality has subscribers in agreement with that realism. Having said that, a reader believing in only one kind of reality might be inclined to object. Certainly, on the surface it seems that only one reality, often called common (or naive) reality, exists. However, when we dig deeper into what we normally consider *reality*, we find that it does have many variations.

In no particular order, here is a short, but representative sampling of the various realisms that might be encountered: Conceptual Realism, Scientific Realism, Conventional Realism, Platonic Realism, Empirical Realism, Critical Realism, and Operational Realism. This list could go on, but as an example it is enough to get across the idea that there is no fixed – one size fits all – reality.

The events in our lives manifest themselves in two distinct, but connected, ways. The natural world of phenomena, for the most part, is objectively presented; however, our interpretation and contemplation of these events is subjectively influenced and individually unique. Neither objectivity nor subjectivity can be avoided and together they determine our personal version of reality. When the objective experience - the phenomenal events common to all - and the subjective interpretation meet certain cultural and social standards, then the individual is said to be normal within the constraints and conventions common to an individual's cultural and social environment.

Simply stated, reality has a certain amount of variability, which depends on such things as individual capabilities, knowledge, social environment, inclination, and so on. To illustrate this point, here are examples of two very different realities.

In the Kalahari Desert of southwest Africa, reside the native Bushmen (San People) who, from the point of view of environment, technology, and formal education, live very primitive lives. Indeed, until recently, these people had lives not far removed from what we commonly call stone-age humanity. At the other extreme are the typical young urban dwellers, perhaps living in Los Angeles or London, who are constantly connected to the latest technical gadgets, especially those starting with 'i'.

These assorted technical accoutrements tend to be constant companions of urban living in general and youthful urban living in particular. On the other hand, the *advanced* technology of the Bushmen is fire generated by wood friction and a bow with arrows. These individuals live in habitats about as far apart socially, environmentally, and technologically as is possible these days.

However, these desert dwellers, although technologically primitive, have an intimate alignment with the natural world, which includes not only physical objects, but also includes a strong spiritual sensibility and an awareness of the integrated wholeness of existence. That is, the San People have knowledge of universal wholeness and the *Unitive Principle.*

In contrast, these urban dwellers, technologically over equipped and spiritually deprived, are mainly occupied with technology and a certain social behavior mediated by current fashion. Put another way, the San People are fully integrated into the natural world, while the young urban dwellers are fully integrated into the world of artifacts and superficiality.[154] It is not hard to imagine that each have a very different view of reality. What one is missing, the other has in abundance.

To be sure, we do not have to go to such extremes to find variations in reality; each of us has a personal view that is never exactly the same as that of another. For example, two people viewing the same rainbow will probably have different interpretations of the event. A meteorologist, for example, will look beyond the phenomenon and see not only the rainbow, but will reflect on the physical causes of rainbows. On the other hand, an artist will most likely see the esthetics of the rainbow. Which interpretation is the *real* one? Of course, both are just as real for each person.

Reality is partially based on our assumption and belief in veridical perception. That is, perception based on or that coincides with what we judge to be real. As the observer's knowledge becomes more complete, in the sense of increased insight into the observed structure and properties of the objects and events experienced, then the observer's reality is modified accordingly. Of course, the object itself does not change; only the observer's image and knowledge of the object is changed. Put simply, seeking knowledge and increasing understanding modifies individual reality and enhances a person's worldview.

---

[154] Unfortunately, the government of Botswana has re-located many of the Bushmen, and their old way of living is quickly fading.

This is somewhat circular but what we perceive and understand depends on our worldview, while, in turn, our worldview depends on what we perceive and understand. Each of us has a private worldview that is unique and overlays the world as revealed by our senses. Such private realities are not problematic as long as they stay within the boundaries of cultural norms and conventions.

Some say that technical knowledge blunts the edge of natural beauty and dulls one's sense of the wonders of nature. This is a misconception. For example, knowing something about what generates a rainbow can and will enhance the observer's appreciation of the wonder of this phenomenon. It is nature bringing together the necessary concomitant structural composition at the right time to produce a thing of beauty.

*   *   *

Perceptual consistency is fundamental to our notions of external reality, and this consistency is achieved through the continuous correlation of active sensations with past sensations. This process is more than the mere collection of data but includes processing the data into information and information into knowledge. Although the stream of sensations seems continuous, our experience of these sensations is episodic in the sense that we consciously perceive intermittently and with variable intensities.

The concrescence or the objective reality of an external object is most strongly determined through perceptual persistence and consistency. Multiple encounters with the same or similar objects reinforce the concept of the existence of a particular object and in this way, the external world as projected into us is integrated with our sense of self to form the totality of our reality.

All versions of reality, at least those considered rational, are built on a fundamental and unalterable basis of all realities. This basis of reality includes, for example, unalterable and unavoidable components of reality such as the irrepressible force of gravity, the limits to biological existence, biological necessities such as sustenance, and so on, and this basis necessarily forms the foundation of all variations of reality, even those that are not completely rational. Given this unavoidable basis or foundation, one might question whether these *variations of reality* are nothing more than mental exercises. However, there can be no doubt that since we pattern our lives on our personal worldview, these variations must be considered truly a part of our personal reality.

On reaching adulthood, we are, for the most part, set in our ways and have established a reality for ourselves, which will influence all our activities and opinions. Although it is possible for a person to modify their worldview somewhat and suppress certain behaviors or features of their reality, a full escape is probably not possible. At the innermost level, our personal reality is deeply ingrained such that changes are difficult or even impossible. We can change cosmetically or superficially, but just as the topiary designer may change its outer appearance, the bush is still the same bush, similarly, the person's basic nature remains.

{**Side Comment on Apple**: When the Apple Company began using the lower case 'i' as part of their product naming conventions, they consciously knew that this naming convention would appeal to those who consider themselves as different – in the sense of feeling superior - from most of humanity. In fact, even prior to this new naming convention, Apple advertisements from the very beginning – mirroring Steve Jobs extreme self-centeredness – targeted those who think themselves special and a cut above the masses and looked down on us ordinary folk who use Windows. The timing for this reference to the 'i' was appropriate for the egocentric or the - *it is all about me* - mindset of the *selfie culture* that is now highly prevalent in this country and around the world. If you walk by an Apple store in just about any mall you will find it full of persons just playing around with various accoutrements with apparent fascination, and if a new product or upgrade is announced a techno-frenzy is bound to follow.}

Science and Reality

"The objects of physics are thus, in their connection according to law, not so much "signs of something objective" as rather objective signs, which satisfy certain conceptual conditions and demands."[155]

In general, prior to the development of quantum theory, objects of scientific study were limited, for the most part, to those existing in the macro-world. That is, classical science was mostly concerned with the world of everyday experience, which is, for the most part, directly observable.[156]

However, with the development of atomic and quantum physics, science moved into the micro-world where direct observation of the phenomena was no longer possible. The micro-world is the world of the unvisible and gives up its secrets only indirectly and reluctantly through subtle hints. Studying the micro-world is somewhat like determining the shape and size of a rock tossed into a pond by examining the ripples.

Additionally, atomic and sub-atomic physics has an interpretive problem in trying to translate the postulated phenomena of the unvisible micro-world into the language of the visible macro-world.

The most fundamental and advanced science still requires discussion and explanation in common everyday language and this need has revealed certain linguistic inadequacies. That is, language is adept at describing the macro-world from a mechanistic, cause-effect, and deterministic point of view, but as the quantum micro-world apparently has none of these attributes, attempts to describe the micro-world are difficult and never completely successful.

---

[155] Ernst Cassirer, Substance and Function. New York: Dover Publications, 1953, 305.

[156] Classical studies of electricity and magnetism were of micro-phenomena but relied on visual macro-effects for interpretation and explanation.

Scientifically probing the unvisible microscopic world reveals only reflections or shadows. These are the *objective signs* mentioned in the above quote, which are revealed to us only through amplification and extrapolations of signals from the unvisible world. This process requires a great many complex steps before reaching a level *observable* and interpretable. Furthermore, these *signals* from the unvisible micro-world lend themselves to more than one interpretation and, therefore, more than one theoretical model. Consequently, any preference for one model over another almost certainly has subjective elements based on personal inclination, motivation, and so on.

Because of limited linguistic capability and variable reliability of our sense organs, there will always be a separation between our scientific knowledge and the things-in-themselves. Nevertheless, science has been remarkably successful in its attempts to overcome our sensual limitations and progress beyond sensory phenomena in an unending quest for ultimate reality.

Science is both a discipline of investigation and explanation. Investigation is the objective part, or at least should be, while the explanation part always contains some portion of subjectivity that is introduced either unintentionally or intentionally by the explainer's worldview, background, agendas, and other modifiers that today even include social and political influences.

The so-called natural laws that science strives to discover and codify are at the fundamental level statistical truths. That is, at the micro-level quantum behavior is observed – which means that behavior of matter at these levels is probabilistic.

Essentially, science is a human process that simplifies the laws of nature. We know enough about these laws of nature such that we are able to build scientific theories to explain our partial knowledge of nature, but these simplified versions are incomplete, and we only have veiled or shadowy clues about the true laws of nature that lie hidden.

Put simply, modern science deals with laws which we tend to see as universal truths but are actually anthropically defined relations produced by human minds, however, to be sure, the laws of true science are formulated with rigor and are required to follow long established procedures.

Influencers of Perception and Reality

The preceding discussion on perception was about the objective part of perception, while reality was shown to have some personal variability. That is, the perceptual process just discussed is the *objective* part of perception and avoided the *subjective* influences that modify and augment our perceived reality.

However, the fact is that personalities as well as our social and physical environments have unavoidable influences on perception such that the subjective elements within perception can only be minimized, never eliminated, and as perception is influenced, inevitably, so is our reality.

These modifiers and influencers fall into the usual two broad categories of the *social* and the *individual*. Those within the social category are the common modifiers and influencers within global society as well as in a person's local social environment, while the individual modifiers and influencers are internal factors such as education, personality, worldview, inclination, and mental state at the moment of perception. Of course, the social has an influence on the individual, as well as the other way around, such that the boundary between the two is indistinct and fuzzy.

Mostly, we are not the cause of what we observe; instead, we are the interpreters of what we observed and as interpreters we cannot avoid ourselves such that even the most objective investigations are not exempt from personal and social influence.

Because of scope and space limitations, the following brief snippets consider only a few of the many potential ways perception and reality are modified or influenced. However, I do think they are enough to make the point that the objective is always influenced by the subjective – in some people more and in others less.

Recollection

"The truth is that memory does not consist in a regression from the present to the past, but, on the contrary, in a progress from the past to the present."[157]

"When I was younger, I could remember anything, whether it happened or not. But as I grew older, it got so I only remembered the latter."[158]

Memory is the recollection of an absent object or past event. That is, the memory of an object not immediately present, but was present in the past or a prior event.

It is easy to see that memory is an indispensable component of the process of perception. Without memory, every sensation would be unique and knowledge beyond the active sensations would not be possible. There would be no prior associations and no experience. Without the ability to recall past experiences, we would be unable to make comparisons that in turn lead us to decisions and action. Put simply, without our accumulated memories life as we know it would not be possible.

However, a recollection is more than just recalling a past memory; it consists also of a reinforcement and possible modification of the recalled memory. We re-think the past memory and update it through blending the old memory with new content. As the word indicates, to recollect is *to collect again* and this aptly describes what happens when current perceptions awaken memories.

As the above quote attributed to Mark Twain humorously put it, there is also the question of the veracity of the recalled memory. Time and other factors tend to corrupt our

---

[157] Henri Bergson, Matter and Memory (New York: The MacMillan Co, 1919), 319.

158 Attributed to Mark Twain.

memories. As we all know too well, over time, our memories tend to become vague and fuzzy, which increases the possibility of unintentionally corrupted recollections. On the other hand, sometimes memories, especially unpleasant ones, might be purposely altered. Either way, the veracity of memory is always suspect.

Recollection augments our current perception by association with recalled similarities, which provides a sense of familiarity. On the other hand, there is the apparent random and capricious recollection, such as a piece of music, which pops into our consciousness for reasons quite unknown. There is no explanation of how this *popping into consciousness* happens, but it does provide a clue about the hidden complexity of the connections between consciousness and the subconscious. There also seem to be exceptions such as the notion of synchronicity.

For the most part, memories are contiguous with the interests of the individual. This contiguity allows the individual to function normally. That is, normal functioning requires that the past, as remembered, correlate positively with current perceptions.

Language

What is language? A simple answer is - it is how we communicate. Language, for humans, is a combination of spoken words, written symbols, facial expressions, and what is commonly called *body language*. When communicating, and we are always communicating in some manner either consciously or subconsciously, we use these elements of language in various combinations and with varying degrees of emphasis.

Cultural content is deeply embedded in all languages and the two cannot be separated or completely eliminated. Even when it is commonly assumed that two cultures speak the same language, each culture has specific language nuances, some subtle and others not so subtle. One only needs to recall the well-known aphorism that *"England and the USA are two countries separated by a common language."*

Language is the association of abstract objects (words written, spoken, and thought) with concrete objects (physical things), abstract objects (mental things), and actions (events). That is, words are conceptual symbols that represent things, thoughts, and actions that only have meaning within the context of a particular linguistic convention. This convention is, in general, not a one-to-one correspondence between words and things, but a one-to-many correspondence such that a word can and often (usually?) does have multiple meanings. Consequently, we must rely on sentential context for the intended meaning.[159]

To complicate matters, as languages evolve words often lose their original meaning but continue to be used. For example, consider the words *sunrise* and *sunset*. Although still capable of conjuring up delightful images, most of us know that the earth rotates and that the sun only appears to *rise* and *set*. Another example is the word *awful* that seems to have reversed meaning somewhere over the years – it appears that it should refer to something wonderful as in *full of awe*, not something bad or disagreeable.

As is well known, language variations are often based on regional and social conventions. However, less recognized (or just less discussed), is the fact that language is also a gauge for personal judgment. How an individual uses language is an indicator of their education, personality, social situation, and worldview. That is, richness of vocabulary or the lack of richness and the excessive use of slang, colloquialism, or profanity

---

[159] For example, one of my favorites is the English word run, which in a particular Random House Webster dictionary included 179 usage definitions!

are personal indicators that say much about ourselves and play a key role in determining how we are perceived by others.

It seems that today, there is much less thought given to how we use language. A few decades ago, it was common to hear quotes from classical literature and poetry, but, today, most often all we hear are catchy phrases from popular movies or television sit-coms that usually do not have much depth of meaning. It is an unfortunate truth that jargon, slang, colloquialisms, and obscenities seem to be what a lot of people consider proper language usage. That is, individuals will purposely adopt certain language conventions as well as affectatious modes of dress and overall appearance in attempting to identify with and be associated with certain social or cultural groups.

{**Side Comment on the Power of a Single Word:** The *Nicene Creed*, first published in the year 325, evolved over time and by the 15th century, the Western Catholic Church, and the Eastern Orthodox Church had different versions of this creed. The Catholic version contained the Latin word 'filioque', which translates as 'and from the son.' The Orthodox version of the creed did not include this word. Consequently, in an age of strong religious ardor, this difference was enough to cause a rift between these two Christian branches such that, in 1453, when Constantinople, the center of Eastern Orthodoxy, was under siege by the Islamic Ottoman Empire, the Christian West would not send aid to the Christian East. The result was the capture of Constantinople, and the history of Europe and Asia were changed forever. Never underestimate the power of a single word!}

{**Side Comment on Ambiguity:** Some say that the legal profession invented ambiguity. Whether or not this is the case, ambiguity is a useful and sometimes subtle tool for 'hedging one's bets' by appearing to state a position without really committing to a particular position. They say that in ancient Greece the Delphi Oracle (aka Pythia, possibly dating from about 1600BC) was famous for making such statements. Anyway, here is a more recent and obscure example. On June 22, 1876, when

LtCol. George Custer was starting on his ill-fated journey to the Little Bighorn River, General Terry told Custer 'Do not get greedy, save some Indians for us.' Custer's reply was 'No, I will not!"

{**Side Comment on Denotation and Connotation**: For most of us, these two words are not likely to be commonly used on a daily basis, but they are two words that should be understood with an awareness of the important distinctions between the two. First, denotation is a reference to what will be called the dictionary definition of a word and is mostly objective, while connotation is the subjective or emotional definition or meaning of a word. That is, many words have a dictionary definition as well as an emotional or socially defined meaning that should be considered in order to avoid misunderstanding or worse. For example, the word *cheap* has a dictionary definition that essentially means inexpensive, but its connotation is aligned with *inferior quality*. This is just one example of the many nuances and idioms that are commonly encountered in ordinary everyday discourse. Another example is the different nuance between *stubborn* and *determined*.

Of course, these days political correctness has taken this dichotomy to the extreme where connotation for some can cause an aggressive response. For example, while writing this in the late spring of 2021, there are ongoing acts of vandalism and lawlessness that are going on in many US cities that are called mostly peaceful protesters by some and domestic terrorists by others when it should be clear that the wanton destruction of private and/or public property is beyond protesting and warrants a label that fits the actions, not a label that attempts to hide the true nature of these action by lumping them together with legitimate and peaceful protests.}

Language and Logic

Our language limits our investigative capabilities, and it also seems apparent that our logic is not the same as the logic of nature or of the Mind. Our ability to conceptualize is strong but limited such that we do not really know how deeply we are able to investigate and understand nature. That is, there is no absolute scale or universal measure that we can make that would indicate the depth at which we understand nature or would provide any type of standard scale such that we could measure our intellectual capabilities universally. We arrogantly flatter ourselves by using terms like theories of everything and conjuring up fantasy notions about the origins of the universe, but what we really know, compared with what is universally possible – there is no way of telling.

At the common level, language is used to discuss events and things of the world, while at a higher level, language is used to discuss language itself and this language about language is a level of self-reference and is sometimes called meta-language.

That is, language symbolically conceptualizes things and events of the world, and it is the richness of our language that facilitates our ability for self-reference. Self-reference is a capability that (we believe) distinguishes human cognition from other live creatures (or any machine). This means that the logic of Mind is not the two valued (true or false) logic of the machine. No machine is self-referencing and most likely, cannot be. There exists no machine algorithm for self-reference or any algorithm that has given rise to imagination and creative thought - and certainly not for one of the most human and mysterious experiences sometimes called intuition or a *leap of faith*.

Put another way, science in its purest sense, literature, and the true arts are not the result of algorithmically driven processes. The working of nature seems to be of a kind that transcends the logic that we are able to implement in our machines such that we may never know these hidden methods of nature.

Sensitivity

When we say that a person is sensitive, it commonly has either of two meanings. It might mean that this person is empathetically sensitive to the feelings and situations of others or is a statement about a person who is easily offended, easily embarrassed, or otherwise emotionally fragile. The problem is that it is easy to confuse these two when someone refers - rather ambiguously - to someone as being *sensitive*.

Then there is also a sensitivity of a third type, this is the sensitivity of awareness or insight. For example, in the case of perceptual sensitivity it means awareness of subtleties such as body language and facial expressions, not just the spoken word. How you respond to others and how others respond to you is often determined by observing and understanding the less obvious and more subtle differences and similarities. This third type is what enables some people to quickly make character judgments while it is the lack of this third type that often labels someone as gullible or naïve in the sense of being a poor judge of character and easily fooled.

Explanations

At first, explanations may not seem closely related to perception or reality; however, since the general purpose of an explanation is to transfer information there is a clear connection between explanation and how something is perceived or understood. Anyway, we spend a lot more time explaining than we imagine and probably think less about the quality of these explanations than we should. There can be no doubt that the quality of an explanation has considerable bearing on how we understand something and that in turn affects us perceptually.

The motivations for explanations seem to have two main purposes; they are intended to instruct (inform), or they are attempts to persuade (influence) – or both. Either way, it is possible to divide explanations into three common types: *example*, *labeling*, and *descriptive*.

The *example* type can be further divided into two subtypes, which will be called *like-example* explanations and *specific-example* explanations. Explanations of the *like-example* subtype use examples of objects with similar properties, but these objects are different from that being explained. This type is used frequently when attempting to clarify an event or object where a known specific example does not exist, is unknown, or not yet experienced. For example, it is said that rattlesnake meat looks like fish and tastes like chicken. What this type of explanation really does is use familiar examples to create a sort of sensory boundary around the object of the explanation and provide a sensory image based on the similarities. Although useful in everyday situations, this type of explanation definitely lacks accuracy or specificity.

On the other hand, the *specific-example* subtype of explanation attempts to explain using examples of the object of explanation. For example, a small clock attached to a wrist by a small band uses specific features of this type of clock. The obvious difference is, rattlesnake meat is neither chicken nor fish, but a wristwatch is a small clock and attaches to a person's wrist with a small band.

The *labeling* type is the weakest of the three. For example, while shopping in a grocery store, I queried a store clerk about an unfamiliar fruit and the clerk's response was to provide the name of the fruit. Obviously, simply naming the fruit, or anything else, is not an explanation.

The *descriptive* type, as the name implies, involves a list of characteristics (properties), and can be broken down into two subtypes of either *operational* or *functional* explanations. The *operational* explanation might also be called the *recipe type* explanation. Although helpful, a list of instructions really does not explain much at all, it merely provides a series of steps. For example, instructions for making a cake tell us little about the cake itself. On the other hand, the *functional* explanation is one that includes the *why-how* or the *cause-effect*. That is, it is not just a list of instructions, but a true explanation, which provides reasons or justifications for the actions.

The point is that the quality and manner of explanations directly influence our perceptions, especially when, as is often the case, the explanation is an attempt to influence the recipient. Such explanations are common in commercial and political advertisements. Both types of advertisements use misleading statements and incomplete information in their attempts to influence opinions. That is, they only pretend to be informational, while actually attempting to influence rather than inform. Errors of omission, purposeful or otherwise, are a common way to make both products and politicians seem more favorable. For example, product commercials frequently use statements such as *15% less* or *doctors recommend* which are meaningless without knowledge of the alleged referents.

These and similar *context-less* statements are common in product advertising, while in politics, we frequently hear vague and broad statements about *making things better* or *helping the middle class*. Here, the omissions are usually the specific ways and means of accomplishing what is promised. Such statements never really explain what the actual process of improvement might be or how things will get better.

Without doubt, the producers of these misleading statements are experts at obfuscation, ambiguity, and so on. We should remember that manipulation of our perception is the business of advertising, and they are particularly good at their business. All advertising has a visible agenda, and it should be assumed there is also a hidden or unvisible agenda.

{**Side Comment on Narrative Explanations**: It should be noted that certain sciences essentially rely on narrative explanations based principally on observation and not on experimental actions and results. Examples include historical geology, paleontology, astronomy (cosmology), evolutionary biology, and so on. Techniques used by these disciplines include such vague statements as '*adequate causal mechanisms*' and '*best approximation of the true casual history*' which sound much like guessing. Also, they never explain how *adequate* or *best approximations* are determined and by whom? The fact is these types of disciplines frequently suffer from underdetermination, which is a euphemism for lack of precision and evidential facts.}

Attention

"Times flies when you're having fun." -Anonymous

As the above quote states, and as all have experienced, our state of mind influences the flow of time and this in turn influences our perceptual awareness or perceptual receptivity. This influence of attention on perception and the flow of time is both obvious and subtle. It is obvious because all of us have experienced varying degrees of attention as well as varying rates of the passage of nature. On the other hand, it is subtle because it occurs more often than we suspect and with greater influence than we suspect (or would like to admit).

If the active sensation indicates a peril or some possibility of danger, our attention is immediate, and depending on the situation the response is either a spontaneous reflexive action or an action based on considered alternatives. At the other extreme is the state of boredom that results from the lack of stimulation, interest, and so on. In between the high intensity levels of attention and the lower boredom like levels of attention is the middle ground of sensations that are of sufficient intensity (interest) to warrant appropriate attention.

What then is the middle ground for optimal attention, which maximizes the attentiveness and efficaciousness of the perceptual process? The answer, like most answers about human behavior, is the not-very-useful answer of *it depends*. Attention depends on the individual and the environmental situation such that the maintenance of optimal attention is often problematic and inconsistent. Furthermore, each person has individualized inclinations and motivations that determine their level of attention within a given situation.

Perceptual Veracity and Other Tricks

"Note that in every illusion what is false is what is inferred, not what is immediately given."[160]

Most of us believe that the perceptions we experience are reliable and trustworthy. However, if we take time to consider sensory reliability, it does not take long to realize that our senses have considerable variances in both reliability and trustworthiness. If we rank the senses in order by trust and veracity, it seems that tactile would rank first and auditory last. The aphorism *out of touch with reality* speaks clearly about how important touch or the tactile sense is to our concept of reality and sense of normalcy. There is something special about the sense of touch, which gives it more credibility than the other senses. On the other hand, have you ever heard anyone say - *Hearing is believing*?

Since, like most subjects discussed herein, it is possible to discuss perceptual veracity at great length, only visual perception will be briefly considered. Although less reliable than the tactile, there is no doubt that, for normal functioning, vision is our most important and complex sensory capability.

---

[160] James, 1910, 86.

Of the possible visual tricks, *illusion* is probably the most common. An illusion is corrupted visual perception caused by erroneous interpretation of sense data and/or by incomplete data. Visual illusions are familiar to anyone who has seen a *magician* perform. Magicians, or more correctly *illusionists*, create visual illusions primarily through misdirecting and anticipation. The illusionist attempts, and usually succeeds, in making us *see* what is not there, and *not see* what is there.

Because we grow accustomed to certain visual patterns, we sometimes *see* these patterns even when they are not actually present. Through anticipation and habit, we often do not notice what is missing or in error. For example, when reading we unknowingly *fill in* missing words or correct erroneous words. This type of illusion is sometimes called the *proofreaders* illusion and is why, even after extensive editing and reviewing, errors still seem to find a way [Sorry if you have found any].

Then there are the visual distortions known as *mirages*, which unlike illusion, result not from erroneous interpretation of sense data, but from erroneous or corrupted sense data. The *water* seen in the desert or on a highway on a sweltering day is invalid sense data that mimics the appearance of water. However, in this case, do not blame your senses; it is just one of nature's tricks.

Since it can be the source of much trouble, one final visual trick will be considered. This is the condition of perceptual misdirection known as *target fixation*. Many accidents have occurred and will continue to occur because of this problem. This condition, common among long haul truck drivers and military aircraft pilots, is a perceptual mistake that almost all have made at one time or another.

For example, when a bicycle rider is trying to maneuver through a narrow passage, such as a gateway, there is a tendency to look at the side of the opening instead of looking at the middle of the opening, and since we all tend to follow our eyes, this can cause the rider to unintentionally turn toward the side of the opening and, unless very lucky, crash. The results for a bicyclist may be only a bruise or two, for a truck driver or pilot the results are more consequential. The lesson is, look at where you want to go, not at what you want to miss.

Mental Practice

It has been well established, that repetition of a motor action eventually creates a *short cut*, which allows us to react quickly and avoid the delay of mental processing normally required prior to physical movement. When motile functions are repeated, these actions eventually automate and require only a negligible delay to respond to a triggering event or situation. This procedure, perhaps best known to athletes, has the common name of *muscle memory* and is part of all athletic training.

What is less known is that even thinking repetitively about an action can influence the actual execution of this action. As with muscle memory, such mental practice has been recognized as enhancing the performance of athletes. For example, the technique of mental practice is common among golfers who mentally rehearse each shot. In fact, it has also been demonstrated that just about anyone can use mental practice to advantage, and its effectiveness has been shown in occupations as diverse as musicians and surgeons.

Repetition, physical or mental, may be thought of as an inductive process that never quite reaches an end, and, in this sense, is always evolving or improving. Each occurrence is a similarity, not a replication. Something is always different such that we dynamically accommodate the differences and anticipate the similarities.

Sometimes this has been called 'reflex action' that is capable of almost instant action such as catching something accidentally let slip. It is an automated mind condition or capability that is difficult to describe or understand such that it is difficult to understand how it originates. Colloquially, such terms as 'having one's head in the game' or 'momentum' are used to describe this mental state of exceptional and reflexive performance. Also, it is fragile in the sense of being easily lost through distraction or some type of interruption, such that practice is required to obtain this capability and discipline is required to maintain it.

It should be noted that improvement through physical practice has a long history. For example, it is known that the ancient Roman gladiators had practice arenas where they honed their various skills. More recently, it has now been recognized that senior citizens are still capable of cognitive retention and even improvement through mental practice.

Or should it be called *mental exercise*, just as muscles deteriorate when unused, mental capability also needs to be exercised through activities and games that provide mental challenges. These games and activities need to be more than just recall exercises such as crossword puzzles, they need to exercise problem solving and inferential thinking as well.[161]

Sensory Pollution

"Put a man into a perfect chaos of phenomena, sights, sounds, feelings; and if the man continued to exist, and to be rational at all, his attention would doubtless soon find for him a way to make up some kind of rhythmic regularity, which he would impute to the things about him, so as to imagine that he had discovered some law of

---

[161] Certain problem-solving computer games such as *Microsoft Freecell* provide better mental exercise than games that rely mainly on recall and luck such as most solitary games and crossword puzzles.

sequence in this mad new world."[162]

This quote, from a work published in 1885, seems very modern. The author was living in Boston at that time, and, as anyone who has ever been to Boston can attest, the chaos of 1885 that inspired this comment has surely increased greatly since then, and given Royce's rather bucolic background, this sensitivity to the chaos of Boston is not surprising.[163] Of course, this is not just a Boston phenomenon, but because of increasing urbanization and industrialization this is exactly what we now experience in every urban environment.

The 20th century was the century of noise where the development of wireless technology first introduced the radio, which became ubiquitous and always seemed to be turned on such that the natural sounds of the world were shut out. Of course, it was not just the radio, it was also the automobile, construction equipment, and so on that generated urban noises at ever increasing levels.

Today, besides urbanization, the development of various technological appliances has enabled and facilitated constant sensory bombardment. Just as urban illumination (light pollution) has all but eliminated the once magnificent nightly stellar spectacle, auditory pollution has smothered the sounds of nature. As if this pollution is not bad enough, we commonly see individuals who seem unable to function without something plugged into their ears, and thereby substitute one form of auditory pollution for another of their choosing.

An adjunct to this increase in sensory pollution appears to be a growing need for over stimulation and constant entertainment that combine to form a closed loop of escalating

---

[162] Royce 1885, 317.

[163] Josiah Royce (American philosopher, 1855–1916) was born, in the small foothill town of Grass Valley, California.

excessive stimulation. This seems especially true in current cinema where special effects and loudness are, at an increasing rate, replacing dialog and story. The use of computer-generated graphics is replacing natural scenery, and this stimulates striving for even more visual stimulation. At the same time, monster sound systems blast ever louder.

Additionally, it should be mentioned that in cinema there seems to be a trend toward apocalyptic scenarios and myriads of super-heroes with outlandish powers. Together these trends seem to indicate a certain fatalistic pessimism along with an excessive desire to escape into fantasy.

Of course, such excesses are not new; this trend of loudness and spectacle has long plagued popular music, where voice augmentation, ear breaking audio, and visual overload are the norm. This constant and extreme sensory stimulation affects perception through over stimulation, which leads to desensitization.

One result of all this pollution is the alteration of our perception of what constitutes entertainment itself. Entertainment has morphed from a form of relaxation, reflection, and even edification to a form of extreme exaggeration and sensory excess that is increasingly devoid of meaningful content.

{**Side Comment on Self-expression:** Along with all this sensory excess, it seems that self-expression has increasingly taken the place of achievement. This is seen especially within the realm of popular entertainment where it is becoming more common for someone to become *"famous for being famous"* such as those appearing in (so-called) reality shows. It seems that a culture of posturing is rapidly replacing the culture of correct behavior, personal integrity, and accomplishment. Examples include the phenomena of affecting a certain fad and pretend accomplishments like walking to cure something, but really to get a t-shirt and a selfie to post on social media. These days we hear a lot of talk about *raising awareness*. I find it difficult to consider simply raising awareness as an accomplishment unless it is followed up by some actual action.

There is also the trend that seems to be attempting to place everyone on the same level, which is essentially an attempt to push down high achievers as opposed to bringing up low achievers. Education seems more occupied with lowering standards than raising the achievement levels of all students. Today students do not quote Shakespeare or other classical writers and poets, they quote lyrics from pop songs and dialog from cinemas. We have also forgotten that equality, under the law, does not mean all are identical or of equal capability, or have the same motivation.}

Judgments (hasty and otherwise)

Hasty judgments are often the result of the desire to make quick, easy decisions, and to avoid the extra effort of a more thoughtful choice. As a result, such decisions are often over-simplifications based on incorrect and/or incomplete information.

Judgments about complex issues are often decided from only a single *fact*, which is accepted as conclusive, while completely ignoring other pertinent issues. This is problematic since basing a judgment on a single assumed fact or belief is weak and sets a fragile foundation.

No doubt, it is much easier to base an opinion on a single piece of supposed evidence rather than consider a more complete and complex justification. However, this type of judgment, besides being fragile, often leads to hasty and invalid generalizations, which, in turn, can result in unwarranted assumptions such as stereotyping as well as the mistake of oversimplification. Susceptibility to this type of error seems most prevalent when strong emotion is involved such as that which is inherent in political, cultural, or religious opinions.

To be sure, being able to decide or establish a position with minimum effort is what makes slogans so popular and effective. Slogans are an attempt to simplify complexity and make it easier to adopt a certain opinion or decide upon a course of action. One is not required to know a lot about any issues or details; one just adopts and *parrots* the slogan. Slogans do not even have to make any sense; they just have to be simple and easy to remember. In the USA, one of the most famous is the slogan from World War II; *Remember Pearl Harbor!* A few years back, there was the political slogan; *Forward, not Backward*, which, although popular, was just an empty phrase.

Personalization is another vehicle for rash and unsubstantiated judgments. This is the case where an individual tends to over personalize the behavior of others based on their own personal choices. For example, one might condemn a certain behavior because of an *I would never do that* attitude. Such personalizing may seem harmless, except that it tends toward negativity, which if left unchecked, can lead to unwarranted prejudice and condemnation.

There is also another type of personalization that is based on personal animosities or dislikes of an individual or group such that no matter what statement or action by these individuals or groups, the response is always negative, rejected as wrong, or even deemed evil.

Another common way a person avoids thoughtful value judgments is through peer pressure or desire for group identification. This group identification (or Groupthink – discussed below – page 170) seems one of the more popular pastimes. For example, over identification with sports teams or individual athletes is rampant and often borders on the fanatical (but then *fan* is just an abbreviation for *fanatic*). This phenomenon is, of course, not limited to sports, it extends to overidentification with celebrities of all sorts, which is why advertisers use famous people or individuals of supposed authority in their commercials. Such inducements are artificial, but effective.

Contextual Displacement

When someone who is not a historian considers what they believe is a historical fact, they commonly do so within the context of present-day society and their own personal views. This constitutes a replacement of context from the social and personal conventions of the time when these events occurred with current attitudes and opinions. The problem is that presenting the past based on current context is both misleading and deceptive.

History is full of injustices that caused many to suffer, but these events should be kept in their original context. To do otherwise is an injustice to those who are wrongly blamed for events that occurred perhaps hundreds of years in the past. To be sure, many events in the past have influence on the present, but this is not the same as carrying forward the events of the past as a reason to perpetuate prior grievances. Carrying forward the sins of the past provides invalid excuses for the continuation of past grievances.

Honest historians approach the past with an open mind and refrain from judgments based on contemporary notions of right and wrong. The objective historian will present what is known along with interpretations that are consistent with the morals and social conventions of the period being discussed.

Unfortunately, far too often history is a revisionist effort that tends toward interpretations and judgements against standards that did not exist in the past. We see revisionist historical interpretations most often when there exists strong political, religious, or other emotional issues that use what occurred in the past as an indicator, cause, or excuse for today's problems, real or imagined.

At this point, you might be wondering what this has to do with perception and the answer is simply that what we believe is a result of perceptual interpretations and what we believe affects our opinion. So, if we blame contemporary individuals for what is now seen as the sins of their ancestors, then our perception of these people is unnecessarily and unfairly negatively biased.

One of the purposes of historical works is to understand the errors of the past, not to bring them forward with displaced context for purposes of placing blame on contemporaries. To displace historical context is not history, but propaganda.

Dynamic Accommodation

Throughout our lives, we are constantly acquiring new knowledge of the world. Even though we execute most of our daily rituals without much thought, in a subtle but very real sense, every time we perform, even a routine task, it is a new experience. Things have changed between each repetition of the experience. All sorts of things have occurred, and the world has moved on. For example, I visited the same coffee house this morning that I visited yesterday, but today's trip was different from yesterday's trip in a number of ways. The traffic was different, different people came into the coffee house, and so on.

Today, like all days, is similar and yet different from yesterday and every other day. The differences are only slightly noticed and the same can be said about the similarities. That is, in the course of our lives, similar-differences and different-similarities are generally overlooked or given little overt thought, if any.

This is because as we go about our daily lives, we dynamically accommodate these subtle differences, and unless we encounter a situation, which is traumatic or in some other way unusual, we tend to ignore subtle changes and in doing so, we dynamically accommodate these changes and receive emotional comfort through familiarity and the maintenance of normalcy.

Intersubjective Agreement

There is a class of agreements within societies that are a part of socially constructed reality and are based on implicitly recognized common agreements. These agreements can be as limited as an agreement between two individuals or as broad as an agreement encompassing whole societies.

An intersubjective agreement is a collaboration and corroboration between individuals, which exists by extension within communities and societies. This type of agreement, in conjunction with the previously discussed dynamic accommodation, is not only essential for normal functioning, but is an essential element of societies and nations.

Being socially and culturally derived, these intersubjective agreements are usually non-binding such that they rely mostly on common understandings, social conventions, traditions, and are not governed by legal statutes. For example, at the level of societies, these types of agreements govern what is considered proper behavior or tacit agreements of confidence in certain aspects of society, such as confidence in the honesty of the election process, facticity of news sources, and the observance of locally accepted social practices. At the more personal level, married couples frequently rely on this type of agreement, although in this case, the agreement might be considered binding.

Moreover, these agreements often serve as a reinforcement or confirmation of believed reality and provide a level of assurance that an individual's personal reality is in line with that of the collective.

Groupthink

Groupthink should not be confused with intersubjective agreements. Groupthink is a common practice within collectives that determines the group's ideological and behavioral dynamics and often allows little or no flexibility as to when someone is permitted to deviate from the group theme or consensus. This is the practice that most of us believe is characteristic of restrictive political systems or militant organizations that allow little or no freedom of individual expression or deviation from official dogma.

However, this is only partially true, groupthink is prominent in societies and organizations that are generally considered open and free, and might include just about any collective association, such as corporations, political parties, religious entities, as well as social, cultural, or ethnic identifications. And even, possibly, within informal groups such as your book club.

Here is a short list, devoid of details, of some of the primary symptoms of groupthink:

- Unquestioned adherence to the group's ideology.
- Unquestioned belief in the morality of the group's acts and beliefs.
- Rejection of any information that might conflict with stated positions.
- Opponents are considered too naïve or ignorant to appreciate the group's policies and opinions.
- Illusion of unanimity within the group and the false notion that silence is consent.
- Direct and often coercive pressure on dissenters.
- Certain individuals take on the role of true-believers, enforcers, and protectors of the group.

Perhaps, after scanning this list, the reader might recognize that perhaps they too are a member of some groupthink association. Or possibly they might believe that this list belongs to *other people* and certainly not oneself! Of course, this list is not inclusive, and not all items listed are necessary for groupthink to exist. Only two or three items may be sufficient for the existence of groupthink.

There is also the question of how many members (or percentage of members) of a group must agree to a commitment such that it becomes a group commitment? Other than the simple '*it depends*' answer, there is no simple or definitive answer. Group hierarchy naturally plays a role as does how vocal are the supporters of a commitment.

Within every group, committee, or social aggregate there exists the group agenda that is public, the individual agendas of group members, and hidden agendas. Additionally, when it comes to group dynamics, all members are never equally weighted: 1) There are always group leaders, implicitly or explicitly. 2) There are always members who are stronger and/or more vocal, and 3) there are those members along for the ride.

It should also be noted that groupthink can apply where there is no formal organization but only a particular lifestyle or other affectation that has followers who indorse and adopt that particular lifestyle or subculture. For example, those who follow a certain fad or affect a certain mode of dress or overall appearance.[164]

Expect the Unexpected

It is an unfortunate truth that within the complex affairs of humans, nothing ever happens exactly according to plan or as intended. Things always turn out differently. This does not

---

[164] Those interested should consult Janis, I. L. Groupthink: psychological studies of policy decision and fiascos. Boston, MA: Houghton Mifflin, 1972.

mean that goals are never achieved, but that within the complexity of life, there is always the unexpected and we never get exactly the intended result. These unforeseen consequences are intrinsic within complex human events and interactions such that they cannot be eliminated.

Even with the best planning efforts our ability to foresee these unplanned events and results is never completely successful, only degrees of success less than 100% are achieved. This phenomenon is so common that it is usually overlooked or not recognized as an intrinsic property of complex affairs.

In part, the unexpected is a result of complexity, and partly because of the lag between cause and effect such that only later do these unforeseen events and results become known. The examples are so many, it is difficult to choose only a few, but here are three that seem representative.

Example 1: It is common within the information technology field that after extensive testing and simulations, major projects often fail when going *live* because of unexpected and unforeseen conditions. The author is personally familiar with the case where a large corporation developed a new stockholder's records system that was developed and tested over many months, but the first time it was used for *real* it failed because the designers had not expected to cut a dividend check greater than $999,999.99 and that when that amount was exceeded – the system failed.

Example 2: This is the well-known phenomena of political upsets, where a strongly (supposed) favorite candidate does not win. For example, the 2016 presidential election.

Example 3: I will leave this one for the reader to select. The essential point is – that within complex human affairs - there is always the unexpected.

Part 2 Recapitulation

Beginning with our earliest sensations of the *not-ourselves* external world, we begin the process of receiving and processing signals from outside of ourselves. During the first months of our lives these signals or sensations seem incomprehensibly random and uncorrelated, however, over time, we begin to make sense of these signals through an ability to recognize repetitious patterns and retain *images* of these patterns in the form of concepts.

Although perception seems like an objective process and can be broken down into objective steps, after the innocence of infancy there are always subjective influences that modify *pure* perception. The sources of these subjective influences are both social and personal. As previously mentioned, society, both globally and locally, influences our perception through convention, peer influence, social expectations, and so on, while at the internal or personal level, our perceptions are influenced by how we interpret our environment, our close associations, and by our unique personal experiences.

There can be no doubt that experience is prior to and necessary for building our perceptual notions of external reality; and prior experience is gained through our sensory capabilities and events. Put simply, it is through the acceptance of perceived objects and events as being something we call real that we know and acknowledge external reality.

Our experience of the world is mediated through reconciliation of active sensations with prior perceptual experience and as the experiences of the past are brought forward for correlation with the current experience, the past is projected into the current experience. The immediate sense data is correlated with memory of similar content. However, this correlation is never exact such that the current observation of a familiar object is always different from the previous view. Something has changed, be it ever so slight.

Because we seek to maintain comfort and normalcy, we tend to seek our reality by the least effort path and avoid the agitation of a difficult path. This applies both to our physical as well as to our mental comfort, since without both there is neither.

The greatest reality is the knowledge of our own existence. It is the *I*, which dominates us all and is the center of our world. Although this sounds *egotistical*, it is not meant in that sense. This is just the reality of existence and how could it be otherwise? The humblest and most selfless person is still the center of their own world (existence). This is *self-centric*, as opposed to *self-centered* and is inescapable. With either greater or lesser degrees of centricity, the personal self is always present in evaluation of sensation and determination of action. Traditionally, self-denial was (and is) the path of all saints of all religions, however, today, such self-denial seems impossible and probably is given the *"it's all about me"* attitude that seems to be pervasive within our society Put simply, with the rise of the *selfie generation*, it seems that self-centered has become the new normal.

Since humans are part of the natural world, should not their constructions also be considered part of the natural world or do these constructions form a barrier separating us from the natural world? We will leave the answer to the first part of the question open and answer the second part in the affirmative. To be sure, technology in its many manifestations tends to close out the natural world both physically and spiritually.

It is only in the remote parts of the world among the most technologically primitive cultures still existing do we find societies that still retain knowledge of the Unitive Principle and are still connected to the wholeness of the natural world. These societies are often deemed superstitious in their relations with nature; however, these societies have spiritual connections with the natural world that technologically endowed societies have all but abandoned and for the most part forgotten.

We live in two worlds, the external material world of events, objects, and sensation; and the internal mental world of concepts, emotion, and thought. Through perception, the reality of the external world enters into ourselves and blends with our mental world such that within the process of life's journey perception and reality are inextricably joined.

Final Comments

"The Red Queen shook her head. "You may call it 'nonsense' if you like," she said, "but I've heard nonsense, compared with which that would be as sensible as a dictionary!"[165]

"All that is known, all that is knowable, is set over in opposition against the absolute being of the object. The same ground, that assures us of the existence of things, shows them to be incomprehensible. Skepticism and mysticism are united in this one point. No matter how many new relations of 'phenomena' scientific experience may teach us, the real objects seem not so much revealed as rather the more deeply concealed by these relations."[166]

"Be it understood then that the writer is fully conscious of the indemonstrable nature of the hypothesis advanced, in so far as it is a hypothesis, and that he admits, in fact insists, that the value of such hypotheses lies not at all in themselves as ontological speculations about an Absolute, but in their success as expressions of the fundamental postulates and purposes of that source of all truth, Conscious Thought."[167]

As promised in the introduction, as we progressed, the subject matter became more speculative and as well stated in the above quotes, the hypotheses herein are unproven, and, for all our current knowledge (and future attempts?), the true essence of existence, most likely, will remain veiled behind the curtain. On the other hand, our knowledge of ourselves assures us that we exist in a way that is indisputable, and, by extension, we conclude that others exist and have thinking minds. Our thoughts are not mere illusions; they are why we know we exist.

Our knowledge is a fractional extraction from the inexhaustible content that lies both within and beyond our

---

[165] Carroll 1871, 133. (I could not resist repeating this quote.)

[166] Cassirer 1953, 303.

[167] Royce 1882, 25.

current capabilities. There is no final theory of everything. All theory is subject to revision or negation and the quest for such a theory of everything seems to have the characteristics of the alchemists' ancient and futile quest for the *Philosopher's Stone*. This, of course, is the legendary substance pursued by alchemists for centuries that was supposedly able, among other capabilities, to turn base metals into gold. Sometimes it was referred to as the *fifth element*.

What has been discussed here is not meant as more than a brief introduction and a mere hint at what lies both within and beyond the boundaries of our present knowledge and capabilities. The shadowy boundaries of today's knowledge will eventually be fully assimilated into the stock of human knowledge. As this happens, new boundaries, currently unimagined, will appear and the cycles of challenge will continue. As the complexity of our thoughts increases, new conceptualizations will arise and bring an understanding that becomes the starting point for the next advances in both knowledge and capability.

At the beginnings of modern science, in the late 16th and early 17th centuries, scientific knowledge was an affirmative adjunct to religious belief, not a destroyer of religious belief, science versus religion only came later, especially in the second half of the 20th century in a trend that has continued into the 21st century.

Perhaps it is worth repeating how the scientific revolutions of the early 20th century – relativity theory and quantum mechanics – constituted profound changes of the first order. It was this new science of the 20th century that made possible the technology of the 21st century, and for better or worse, this technology has an increasing impact on all humanity such that these changes have brought about a new way of thinking about reality and the natural world that is mechanistically inclined and anthropically self-focused.

When considering these scientific and social changes, it is apparent that the more challenging is the new knowledge of matter and its behavior as revealed by quantum theory. With the development of quantum theory, science learned that the stuff of perceived creation was not what they had thought. As nicely stated by Ernst Cassirer (second quote at the beginning of this section), it seems apparent that as science probes ever deeper into the nature of substance, ultimate reality retreats deeper into the shadows.

As mentioned in the introduction, current knowledge of the vastness of the universe and the unexpected strangeness of the micro-world require that we rethink some of our most strongly held beliefs. We need to remember that many of our beliefs originated thousands of years in the past, in the setting of a very small universe, where the earth was believed to be the center of that universe.

We need to keep in mind the vast changes in our knowledge of the natural world that have occurred over two thousand plus years require that we reexamine many of these ancient notions. The estimated expanse of the universe of the Ancients was no more than a few thousand miles (but usually far less distances were actually ever experienced). For example, the known world of the ancient Greeks only extended in the West to Gibraltar and in the East perhaps to the Caspian Sea. To the early Christians, Jerusalem was the center of the world which extended about the same as the Greek 'world.' The whole of creation was the earth covered by a canopy of stars with heaven above and hell below the earth's surface.

Today, the size of the observed universe has 'grown' in our knowledge such that earth and its environs are but a microscopic speck in the galaxy while in turn the galaxy is but a mote in the universe.

Although modern cosmology has many questionable notions, enough of it is established to the point where rethinking our place in the universe is both desirable and necessary. For example, perhaps the effective universe for us is only a small portion of this vast cluster of stars that we call the Milky Way.

At the other extreme of the *size* spectrum, the micro-world also requires that we rethink our ideas about the basis of fundamental substances. This, along with rethinking the structure of the universe and the fundamental basis of reality, has led to thinking about our place in the universe in a different way. That is, different in the sense of recognizing the wholeness of the universe, and the realization that we need to return to a different relationship with the maintenance of proper relations within society and within the natural world.

This does not mean that we should turn away from technology and return to a simple bucolic lifestyle but rather that we need to keep technology in its proper place as a tool producer and that the products of technology are simply tools – no matter how complex an appliance may seem– and should not be revered as the new idols.

In Part 1, the true nature of space and time were identified as convenient notions concerning movement, position, and change. That is, both time and space are mentally constructed relational notions that are so ingrained in our consciousness that we commonly regard them as some kind of stuff in the sense of having some kind of physical existence. That is, as with time –which is just a relational concept – many people have trouble giving up the notion that space is a container of some kind and only exists as relationships between objects.

This identification of time and space as physical is not a problem within everyday common usage and is useful for maintaining normalcy and the comfort factor that goes along with treating time and space as being some kind of stuff. It is only when we contemplate nature at its most fundamental level of our capabilities that the true nature of time and space must be considered.

Part 1 also discussed various scientific theories of the recent past as well as those that are currently being formulated and how these current notions seem to lack the empirical veracity of past theories and the scientific method that was so successful for the last few hundred years, seems to have been diluted in recent decades through the introduction of unscientific notions and value-based content.

Part 2 took on the task of discussing the connections between perception and reality such that one always has influence on the other, and that although there are the objective aspects of perception, the subjective always has influence on perceptions and consequently our notions about reality are also modified with the subjective, either consciously or unconsciously.

Put another way, perception and reality were shown to be inextricably connected such that our reality is modified by subjective modifications to our perceptions that are used by many (most?) to buffer and isolate themselves from certain realities of life and create the false notion that reality can be made different through acts of mostly symbolic posturing and gesturing such as *raising awareness,* which is just an empty phrase and a feel-good notion for the participant that is quickly forgotten and usually no real action takes place.

*   *   *

Is life merely the emergence of certain properties due to an appropriate aggregation of organic molecules or is the basis of life found at all levels and in all world-stuff? Does a grain of sand have life potential? What is in the nature of fundamental substance that lends itself to the creation of a thinking organism with self-referencing and reflective consciousness? What concrete knowledge do we have of the intrinsic properties and capabilities of matter at the most basic levels? These are questions without definitive answers, and although they remain unanswered, as long as humanity survives the quest for answers will continue.

As discussed herein, science has no answers for intimate knowledge. Notions such as purpose, value, conviction, and so on cannot be explained by the products of physical processes and in this domain, science is impotent such that the methods of science do not apply. Science, of course, is a work in progress and is of such a nature that no final solution or theory should be expected soon or in the foreseeable future.

I know of no conflict between science and religion that can truly be said to nullify spirituality and the mystical. Scientific knowledge has now reached a point where it has become less empirical and has moved on to the frontier of metaphysics. Even after a hundred years, quantum theory, by its very nature, is still misunderstood and confusing. For example, the lack of reassuring deterministic causality is still a test of a person's ability to comprehend and lies outside of our comfort zones of reality as commonly experienced. This situation seems to have stymied progress to the extent that today's science seems more interested in social and political agendas than in scientific progress. This is why make-believe notions such as the *multiverse* have been accepted as real science.

Over the last few decades, most of what is called scientific progress is actually engineering and technical developments and not pure science. This is said, not to disparage engineering and technology but to point out that no major scientific advances have been constructed since the development of quantum theory in its various interpretations.

*   *   *

The interaction of just two individuals is a reflection bound in the history of humankind. The very language we use is an expression of the history of our culture. Even though, on the surface, cultures have considerable variance, the underlying structure of all cultures reflects universality, a wholeness that is interconnected so that the acts of humanity constitute a single drama.

Unfortunately, this universality does not mean that we are all the same or that we all want the same things in life. Just as there are diversities of language and culture, there are diverse worldviews and cultural objectives that can be incompatible with each other.

It is an uncomfortable truth that the current human condition, worldwide, makes it necessary that cultural entities be both open and closed. There must be a balance between being open and being closed that protects the integrity of a cultural tradition.

That is, where there is openness, there needs be assimilation into the host culture such that the aspects of the host culture that attracted in the first place are preserved and that openness is not used to undermine the current culture or to stand outside of the culture and create islands of disparate cultures. The infusion of diversity adds value only when it is integrated into the hosting culture.

The last comment will be about a word that has not yet been used. This is the word *hope*. Hope is the hidden strength of humankind and surpasses all other emotions. Hope is the emotion that drives us to continue with considered optimism that carries us through the darkest situations and most desperate hours.

"Knowing what is enough, you will not be humiliated.
Knowing where to stop, you will not be imperiled."[168]

---

[168] Lao Tzu 2005, 44.

Cited Works and General References

Aczel, Amir. *Entanglement*. New York: A Plume Book (Penguin), 2003.

Alexander, Samuel. "Space, Time, and Deity," *The Gifford Lectures at Glasgow*, 1920. https://archive.org/details/spacetimeanddeit00alexuoft>

Bergson, Henri. *Creative Evolution*. Translated by Arthur Mitchell. London: MacMillan and Co, 1911.

_____. *Matter and Memory*. London: George Allen & Unwin Ltd. Ruskin House, 1919.

Bertolam, Orfeu. "The Mystical Formula and the Mystery of Khronos," 2008. arXiv:qr-qc/0801.3994v1.

Bishop, R. and H. Atmanspacher. "Contextual Emergence in the Description of Properties," (2006) http://philsci-archive.pitt.edu/2934/

Bitbol, Michel, "Reflective Metaphysics: Understanding Quantum theory from a Kantian Standpoint," 2010. http://philpapers.org.

Bohm, David. *Wholeness and the Implicate Order*, Routledge Classics, NY, NY, 1980.

_____. *The Undivided Universe*, Routledge, New York, 1993.

Bruno, Giordano. *On the Infinite, Universe, and Worlds*, Venice, 1584. https://adoc.site/download/giordano-bruno-on-the-infinite-universe-and-worlds-a5b31f76133c56.

Capra, Fritof. *Tao of Physics*. Boston, MA: Shambala Publications 3rd Edition, 1991.

Carroll, Lewis. *Through the Looking-Glass, and What Alice Found There* (1871). New York: Barnes and Noble, 2012.

Cassirer, Ernst. *Substance and Function*. New York: Dover Publications, 1953.

Chalmers, David J. "Consciousness and its Place in Nature." S. Stich & T. Warfield, eds, *The Blackwell Guide to Philosophy of Mind* Hoboken, NJ: Wiley- Blackwell, 2003.

Clifford, William K.: "On the Space Theory of Matter," *The World of Mathematics (Volume 1)*, Editor: James R. Newman, New York: Simon and Shuster 1956.

Darwin, Charles. *Origin of the Species* (1859). Complete works are available at darwin-online.org.uk

_____. *The Descent of Man* (1871). Complete works are available at darwin-online.org.uk.

d'Espagnat, Bernard. *Veiled Reality*, Reading, MA: Addison-Wesley 1995.

Dewey, John. "Perception and Organic Action," The Journal of Philosophy, Psychology and Scientific Methods, Vol. 9, No. 24, 1912.

_____, *Art as Experience*. New York: Penguin Group, 2005.

Disney, Michael. "The Case Against Cosmology", arxiv: astro-ph/0009020v1, (1934) 2000.

Eddington, Arthur S. *The Nature of the Physical World,* Newcastle upon Tyne, UK: Cambridge Scholars Publishing, 1927.

Einstein, A. and Pauli, W. "On the Non-existence of Regular Stationary Solutions of Relativistic Field Equations," *Annals of Mathematics, Vol. 44 No. 2*, 1943.

_____, (1915): A. Einstein, *The Foundation of the Generalised Theory of Relativity* (1915): https://en.wikisource.org/

Ellis, George F. R. "Issues in the Philosophy of Cosmology. Section 2.8.1, Misconception 1," (2006) arXiv: astro-ph/0602280v2.

Epstein, P. F. "The Fine -Tuning Argument and the Requirement of Total Evidence", *Philosophy of Science Journal Volume 84 #4*, 2018.

Feynman, Richard. *QED: The Strange Story of Light and Magnetism*. Princeton, NJ, Princeton University Press, 1985.

_____, *The Feynman Lectures on Physics: Volume* I. Reading MA, Addison-Wesley, 1989.

Gilder, George. *Life after Google*, Washington DC, Regnery Gateway, 2018.

Gödel, Kurt: *Collected Works Volume III*. Oxford: Oxford University Press, Eds. Feferman, Solomon, 1995.

Grandpierre, Attila. "The Dynamics of Time and Timelessness," In the Nature of Time: Geometry, Physics and Perception. *NATO Science Series II. Mathematics, Physics and Chemistry* - Vol. 95, Netherlands, Kluwer Academic Publishers, 2003.

Greene, Brian. *The Elegant Universe*. New York: W. W. Norton & Company, 2003.

Hartnett, J. G. (2011): "Is the Universe Really Expanding?" https://arxiv.org/abs/1107.2485

Hawking, S. A Brief History of Time: From the Big Bang to Black Holes. New York, NY: Bantam Books, 1988

_____. "Gödel and the End of Physics," http://www.hawking.org.uk/godel-and-the-end-of-physics.html.

Herodotus. *The Landmark Herodotus: The Histories*, Ed. Robert Strassler, New York: Pantheon Books, 2007.

Holland, John H. *Emergence: From Chaos to Order*. Reading, MA: Perseus, 1998.

Hubble, Edwin. "Effects of Red Shifts on the Distribution of Nebula," *American Astronomical Society – Provided by NASA Astrophysics Data Systems*, (1936) http://adsabs.harvard.edu/cgi-bin/basic_connect?qsearch=edwin+hubble&version=1.

James, William. *Principles of Psychology Vol. 2*. Cambridge, MA: Harvard Univ. Press, 1910.

_____. *Principles of Psychology the Briefer Course*. Cambridge MA: Harvard University Press, 1923.

_____. *The Varieties of Religious Experience*. New York NY: Longmans, Green, And Co, 1917.

Janis, I. L. Groupthink: Psychological Studies of Policy Decisions and Fiascoes. Boston (MA, USA): Houghton Mifflin, 1972.

Jung, Carl. *Synchronicity: An Acausal Connecting Principle*. Princeton and Oxford: Princeton University Press, 1960.

Kapur, Rajev. AI Made Simply: A Beginner's Guide to Generative Intelligence. LA, CA: Rinity Media, 2023.

Kragh, Helge. "The Controversial Universe: A Historical Perspective on the Scientific Status of Cosmology", (2007) https://d-nb.info/101824364X/34.

Kroupa, Pavel. "The dark matter crisis: falsification of the current standard model of cosmology." 2012. https://arxiv.org/abs/1204.2546v2.

Lao Tzu, *Tao Te Ching*, New York: Barnes and Noble Classics, 2005.

Leighton, J, A. "Perception and Physical Reality," *The Philosophical Review, Vol. 19, No. 1 (Jan. 1910), pp. 1-21.* Duke University Press on behalf of Philosophical Review Stable. (1910) http://www.jstor.org/stable/2177636.

Linde, Andre. "The Uniformity and Uniqueness of the Universe," *This Idea Must Die*: Ed. John Brockman, Harper Perennial, New York, 2015.

McTaggart, John Ellis. "The Unreality of Time," Mind: *A Quarterly Review of Psychology and Philosophy* 17: pages 456-473. (1908) http://www.ditext.com/mctaggart/time.html.

Meister Eckhart. *Meister Eckhart, A Modern Translation.* Trans: R. B. Blakney, New York: Harper & Brothers, 1941.

Merleau-Ponty, Maurice. *The Primacy of Perception.* Evanston, IL: Northwestern University Press, 1964.

Milne, Edward Arthur. *Modern Cosmology & the Christian Idea of God.* Oxford: Oxford Press, 1951.

Nicholas of Cusa, *De Docta Ignorantia II (Learned Ignorance II)*, Chapter Three (1440). http://www.jasper-hopkins.info/DI-Intro12-2000.pdf http://www.jasper-hopkins.info/DI-Intro12-2000.pdf

_____, *Vision of God.* https://archive.org/details/TheVisionOfGodByNicholasOfCusa-OphthalmosAploisOrTheSingleEye-/page/n0

Oppenheimer, J. R. "The Oppenheimer Years," *Los Alamos Science Winter / Spring* (1983) http://la-science.lanl.gov/lascience07.shtml.

Pauli, Wolfgang. *The Theory of Relativity*. New York: Dover Publications Inc, 1981.

_____. *Writings on Physics and Philosophy*. Eds. C. Eng and K. von Meyenn, Trans: R. Schlapp, 1994.

Penrose, Roger. *The Road to Reality*, Ny, NY, Alfred Knopf, 2006.

Poincaré, Henri. *Science and Method*. London: Thomas Nelson and Sons (Trans. Francis Maitland, 1918 edition).

_____. "The Measure of Time," *The Foundations of Science (The Value of Science)*. New York: Science Press, 1913. archive.org/details/foundationsscie01poingoog.

_____. *Science and Hypothesis*. New York, NY: Dover Publications Inc., 1952 (Originally published 1902).

Popper, Karl. Conjectures and Refutations: The Growth of Knowledge. New York: Harper and Row, 1963.

Pratt, James. "Realism and Perception," *The Journal of Philosophy, Psychology and Scientific Methods*, Vol. 16, No. 22, 1919, http://www.jstor.org/stable/2940132.

Pseudo-Dionysius the Areopagite, *Mystic Theology*. T. Davidson translation, *Journal of Speculative Philosophy* vol. xxii, 1893.

Randall, Lisa and Sundrum, Raman. "An Alternative to Compactification." (1999) http://arxiv.org/pdf/hep-th/9906064.pdf

_____. *Warped Passages*. New York: Ecco, 2005.

Reichenbach, Hans. *The Direction of Time*. New York: NY, Dover Publications Inc., 1984.

Riemann, Bernhard: "On the Hypotheses which lie at the Bases of Geometry" in *God Created the Integers*, Philadelphia – London: Running Press, Editor: S. Hawking, 2007.

Royce, Josiah (1882): "Mind and Reality," *MIND*, 1882. https://archive.org/details/mindreality00roycrich

_____. *The Religious Aspects of Philosophy*. Boston and New York: Houghton Mifflin Company, 1885.

Russell, Bertrand. *The Philosophy of Logical Atomism*. London and New York: Routledge Classics, 2010.

Sellars, R W: "Panpsychism or Evolutionary Materialism," *Philosophy of Science*, vol 27, number 4, 1960.

Schilpp, Paul Arthur. *Albert Einstein: Philosopher– Scientist, The Library of Living Philosophers*. Ed. Paul Arthur Schilpp. Tudor Publishing Company NY, NY, 1949.

Smolin, Lee. "The Big Bang was the First Moment of Time," *This Idea Must Die*: Edited by John Brockman, Harper Perennial, New York, 2015.

_____. *The Life of the Cosmos*. N Y, NY: Oxford UP, 1997.

Stapp, H. P. *Mind, Matter, and Quantum theory*. Berlin Heidelberg: Springer-Verlag, 2009.

St. Augustine of Hippo. *Confessions of St Augustine*. New York: Modern Library, 1999.

Wang, Hao. *A Logical Journey*. Cambridge MA: MIT Press, 1996.

Wesson, Paul. *Five-Dimensional Physics*. London: World Scientific Publishing, 2006.

Weyl, Hermann. Philosophy of Mathematics a
nd *Natural Science*. Princeton, NJ: Princeton University Press, 1949.

Whitehead, Alfred North. *The Concept of Nature*. Cambridge: Cambridge Press, 1920.

Wilber, Ken. *Quantum Questions*. Boston MA: Shambhala Publications, 2001.

Yourgau, Palle. *A World Without Time*. Cambridge, MA: Basic Books, 2005.

Zukav, Gary. *The Dancing Wu Li Masters*. New York: Perennial Classics, 2001.Index

Index

www.ingramcontent.com/pod-product-compliance
Lightning Source LLC
Chambersburg PA
CBHW070859290526
45795CB00001B/178